The NAMING
of the SHREW

John Wright

B L O O M S B U R Y
LONDON · OXFORD · NEW YORK · NEW DELHI · SYDNEY

Bloomsbury Paperbacks
An imprint of Bloomsbury Publishing Plc

50 Bedford Square
London
WC1B 3DP
UK

1385 Broadway
New York
NY 10018
USA

www.bloomsbury.com

First published in Great Britain 2014
This paperback edition first published in 2015

British Library Cataloguing-in-Publication Data
A catalogue record for this book is available from the British Library.

ISBN: HB: 978-1-4088-1698-1
PB: 978-1-4088-6555-2
ePub: 978-1-4088-2035-3

2 4 6 8 10 9 7 5 3 1

Designed by Libanus Press

Typeset by Hewer Text UK Ltd, Edinburgh
Printed and bound in Great Britain by CPI Group (UK) Ltd, Croydon CR0 4YY

To find out more about our authors and books visit www.bloomsbury.
com. Here you will find extracts, author interviews, details of forthcoming
events and the option to sign up for our newsletters.

A NOTE ON THE AUTHOR

JOHN WRIGHT is a passionate natural historian and the author of the River Cottage Handbooks *Mushrooms*, *Edible Seashore*, *Hedgerow* and *Booze*. He gives lectures on natural history and every year he takes around fifty 'forays', showing people how to collect food from hedgerow, shore, pasture and wood. Fungi are his greatest passion and he has fifty years' experience in studying them. John Wright is a member of the British Mycological Society and a Fellow of the Linnaean Society. He lives in rural west Dorset with his wife and two daughters.

For Bryan, who knows all the names

CONTENTS

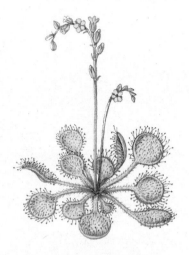

Drosera rotundifolia

PROLOGUE

The beginning of wisdom is to call things by
their right names. (Chinese proverb)

M Y BEST FRIEND AT SCHOOL, PETER, HAD WHAT SEEMED TO ME TO be the coolest father in the world. He owned a model aeroplane shop. Every Sunday we would drive in an ancient Volkswagen camper van to the New Forest in Hampshire, where we would fly his wonderful, expensive, radio-controlled models. Occasionally one would make an unwelcome bid for freedom and fly out of radio range. We would have to chase after it, sometimes for miles.

As well as model aeroplanes and a healthy obsession with rockets and blowing things up, I had a faint interest in natural history (due entirely, it must be said, to a pathetically doomed attempt to impress a young lady). It was on one of those model aeroplane recovery missions that I came across a tiny and very unusual-looking plant. In fact, there were thousands of them, nestling and glistening in the sandy ground. Leaving the fate of the lost aeroplane in the hands of Peter and an increasingly anxious Peter's dad, I stopped to dig one out.

I took the plant home with me and later looked it up in the library. It was, I discovered, a carnivorous plant, a sundew. A pretty enough name, but the one that caught my attention was italicised next to it: *Drosera rotundifolia*. It rolled around the tongue – *Drosera rotundifolia, Drosera rotundifolia* . . .

Obviously just a posh name for the sundew, I thought, but what did it mean? *Rotundifolia* was a simple puzzle: coming from 'rotund' and 'foliage', it means 'round leaves'. And, yes, it did have round leaves. *Drosera* required more study. I learned that it comes from the Greek *droseros*, meaning 'dew', and refers to the tiny, sticky, fly-trapping dewdrops on the ends of the hairs that cover those round leaves. So the name made sense: 'thing with round leaves covered in dew'. Very neat, and more useful, I thought, than the common name. In addition, the book told me, there were others: *Drosera longifolia*, the great sundew, and *Drosera intermedia*, which presumably has leaves whose shape is halfway between the other two (it turned out to be called the oblong-leaved sundew). I was impressed by my discoveries and have never forgotten those three names to this day.

I noticed one more thing. After the words '*Drosera rotundifolia*' there was a single capital 'L'. Slightly red-faced, I asked the librarian (she was way out of my league) where I might find out what it meant. She needed no book: she told me it was an abbreviation for the name of the man who had first described it. He was called Linnaeus.

This eighteenth-century naturalist, she explained, had not only devised the name *Drosera rotundifolia*, he had created the entire naming system for species. But, entranced by Linnaeus's inventions as I was, I could not see the point of them – apart, perhaps, from affording me an opportunity of impressing friends, parents and teachers with my erudition (a noble cause). Why, I wondered, would anyone go to the trouble of inventing puzzling names in a foreign, dead language for plants and animals that already had perfectly good common names, especially considering that the Latin ones were difficult to remember and, all too often, difficult to pronounce?

It was not only the sundew that I found during my New Forest adventures; even more exciting were the fungi. The one that changed everything was an unprepossessing round, charcoal-like lump attached to an ash tree. It was so dead-looking that I could scarcely believe it was a living thing. Appalled that I had no idea what it was, other than perhaps some sort of fungus, I bought the *Observer's Book of Common Fungi* for five shillings, whereupon I discovered that it was *Daldinia concentrica*, a member of the Pyrenomycetes, that shoots its spores from tiny holes in its surface and was very much alive indeed.

My early interest in natural history (thank you Julia, wherever you are) blossomed into a lifelong passion for fungi, and the purpose of Latin names became clear. They are a universal currency across cultures and languages, providing consistent names for both familiar organisms and those organisms that neither have a common name nor ever will. Without Latin names, chaos would rule the science of biology.

I still go to the New Forest a dozen or more times every year, but not to fly model aeroplanes. Now my New Forest visits are to take people on fungus forays. My hope is that my companions will come to love the fungi as much as I do, and maybe, just maybe, reconcile themselves to those much feared Latin names. The latter, however, is not always an easy sell. People find Latin names unpronounceable, unmemorable and unhelpful, closing their ears and eyes to them. I think this is a great pity, because they are, in truth, beautiful, fascinating, often amusing and always useful.

Latin names of a sort existed long before Carl Linnaeus's work in the mid-eighteenth century, but it was he who came to use them consistently as simple pairs (known as binomials). These two-part names put each species in its place: the first word indicates the group, or genus, that the species belongs to (sister species share the same generic name, indicating that they belong to the same genus and are therefore close relatives on the 'tree of life'); the second word – the specific epithet – differentiates the species. Binomials are part of the hierarchy of names developed by Linnaeus: each species nested within a genus, each genus within a 'family', each family within an 'order', and so on.

I impress on people that learning Latin names will enable an ordering of the species in their own minds, because the names themselves reflect that order. Species within a single genus will, on the whole, look similar to one another. And their names often, if not always, tell us something distinctive about them. The slender mushroom *Mycena haematopus*, for example, produces a red fluid from its stem when broken: *pus* means 'foot' and *haemato* 'blood'. *Polyporus squamosus* is a large, scaly bracket fungus, the underneath of which has thousands of little holes in which the spores are created. *Poly* is 'many', *porus* means 'pores' and *squamosus* is 'scaly'.

Since people love stories, I also relate to them the sort contained in this book. A favourite (with me anyway) is the one about the genus *Cortinarius*, which has a dense radiating web of fine fibres covering the young gills. It is called a cortina, from the Latin for 'curtain'. Some have a white cortina, some a blue cortina, and they all go rusty in the end.* Some Latin names are funny without so painful an exposition: few foragers are able to bring *Leucocoprinus brebissonii* readily to mind, but none will forget the name of the stinkhorn, *Phallus impudicus*.

In writing this book my aim is to inspire a delight in Latin names. I tell the stories behind the names, how they were devised and what they mean. Above all, these stories are those of the men and women who created them. Where the complexity of the natural world meets the fallibility of the human mind, tales of triumph and disaster, creation and confusion, honour and jealousy ensue. For the scientists and naturalists who create names, there is much to know, much to learn and so very much that can go wrong.

The arcane rules that govern the coining and fate of Latin names – many of which were laid down by Linnaeus himself – reflect this perfectly. They are of baroque and strangely endearing complexity. There are, for example, no fewer than fifty rules and recommendations dealing with the seemingly straightforward matter of how authors of plant names are recorded next to the species name (as in the L. for Linnaeus after *Drosera rotundifolia*). At the very least, in exploring these rules and explaining the reasons for them, I hope to answer the heartfelt complaint of anyone even slightly familiar with Latin names: why, if they are intended to be consistent, do they change all the time?

Since Latin names wear their hierarchical nature on their sleeves, the story of how they came into being is best understood

---

* The rust-brown spores fall from the gills, eventually turning the cortina the same, final, colour as my MkII and both of my MkIIIs.

against the backdrop of the ancient quest for the 'natural order' of the living world (why species seem to come in discreet but related groups rather than an endless continuum). With Darwin, the natural order became clear: it takes the form of a family tree, the mechanism of which is common descent.* However, the work of placing species within that order is a continuing and contentious challenge. Taxonomy, the classification of the living world, of which naming is but a part, is a vibrant and essential discipline within biology, and the image of aged and desiccated specimens being minutely examined by aged and desiccated naturalists belongs to the past. The living world is orders of magnitude more varied and complex than Linnaeus ever suspected, and even fundamental concepts, such as what may or may not constitute a species, can pose questions that remain unanswered.

Throughout this book I use the term 'Latin names'. Many, perhaps most, scientists prefer the description 'scientific names'. Neither is entirely satisfactory: Latin names are more often of Greek or other non-Latin derivation (they are merely 'Latinised'), while 'mass spectrometer' is as much a scientific name as *Quercus robur* (European oak). 'Latinised scientific names of biological species' is accurate, but hardly convenient.

Since Latin names have a bad habit of changing all the time, there are many, many leftover, out-of-date names. For my purposes, they are more priceless treasure than useless clutter, and I use them

* A natural order is the way in which species are truly ordered, rather than an arrangement of convenience, such as alphabetical order or order by usage (medicinal, perhaps) or by physical characteristics, i.e. the shape of flowers. For most of history the natural order was simply the way God had laid out the world, and common descent, the *true* natural order, was unimagined.

alongside current, correct names indiscriminately and generally without explanation.

The names themselves are the heroes of this book, and I have taken examples from botany, zoology, phycology (algae), bacteriology and more. However, I make no apology for a blatant preference for the names of fungi. I like fungi.

Viburnum opulus

CHAPTER I

WHY *do we use* LATIN NAMES?

I WAS SITTING AMIDST THE CLAMOUR OF MY FELLOWS WHEN THE dark and skeletal form of Dr Parker walked into our classroom for the first time. He turned an unsmiling face to his assembled, and suddenly silent, Latin class and, looking a little like an under-nourished owl, stood and stared. After two minutes of increasingly uncomfortable silence, he exclaimed: 'What a disgusting smell of *boy*!'

This unpromising beginning might explain why I never took to Latin at school. In the third year, Dr Parker called me into his study to tell me that it would be in the best interests of everyone concerned if I took extra maths or geography instead. I am of course being unfair to Dr Parker, a thoroughly nice fellow who had simply hit upon an excellent method of making a useful and lasting first impression. The real reason for my failure as a Latin scholar was that it was hard work and I couldn't see the point of it. No one spoke it, no one wrote it and most Latin texts could be obtained in English translation in the extremely unlikely event I would ever want to read one.

It was during my first Latin-free term that I found the *Daldinia* in the New Forest and my lifelong obsession with fungi and their strange names began. Some thirty years later, I started to share

my enthusiasm with others by leading fungus forays. These are
jolly affairs, when we spend a day searching wood and field for
any fungus that might be found. I name and discuss every new
find and trust that my students will go home with a new appreci-
ation of the fungal kingdom. People are always enthusiastic, and I
happily field endless questions all day long. One particular variety
of question, however, is a little more trying than most.

'I found some trooping pink fairy bonnets in the garden the
other day; do you see them very often?'

So starts many a conversation.

My invariable reply, given through the most pleasant smile
that gritted teeth will allow, is: 'I'm sorry, I haven't the faintest
idea what you are talking about.' Crestfallen and concerned by my
unexpected ignorance, my enquirer will nevertheless persevere:
'Surely you know them, I see them all the time.'

But, of course, I do *not* have the faintest idea what they are
talking about. A trooping pink fairy bonnet could be any
mushroom, at least any that is pink and with an inclination to
troop. I can think of thirty species that fit the name. At least it is
better than 'brown cap', say, which would suit several thousand
species.

'Trooping pink fairy bonnet' is a common name, just like
'sundew', not in the sense of being vulgar (although they can be
that too – the dandelion is sometimes called 'piss-a-bed', for
example), but in the sense of being everyday names for things.
Unfortunately the everyday names of plants, fungi and animals do
not travel well. For example, what is now usually known as the
guelder rose (*Viburnum opulus*) is the Lancastrian's dogwood, the
Kentishman's gatterbush, the Somersetshireman's mugget and
the Gloucestershireman's king's crown. I rather like the name

[2]

they give it on the Isle of Wight – 'stink tree': a highly appropriate name, as I know to my cost, because the berries smell (and taste) of sick.[1]

Things have settled among the plants reasonably well in Britain and many other parts of the world, where more-or-less standard countrywide names for native plants have been encouraged by guidebooks and modern methods of communication. Local names are disappearing into history, lamented only for their charm, not their utility. However, 'guelder rose' is not used throughout the English-speaking world: the plant is also known as water elder, European cranberrybush, American cranberrybush and snowball tree. And, of course, the guelder rose does not insist on living among English speakers. In Spain it is *mundillo*, in France *boule-de-neige*, in Germany *Schneeball*, in Russia калина and in Korea 불두화. There may be a thousand names for this one plant.

Animal names are less variable than those of plants, and all English speakers will know what you mean if you say you have been bitten by a dog or eaten a salmon. But variation still exists. Not everyone would understand if you said you had shot a black-neb (crow) or tripped over a foulemart (polecat). And the problem with names from other countries and cultures still remains.

Mushrooms and other fungi suffer most peculiarly in the matter of common names because they seldom have one. With interest growing in wild fungi for food, this absence has encouraged people to make names up. I am afraid I have done so myself when someone has insisted that I provide a common name for a mushroom they have found. I come up with random gems, such as 'copper knight', 'ash shield' and 'clustered elf ear'. People always seem happy enough to hear my dubious inventions, even when I explain that I might have provided them with the wrong one and that the fungus might not actually have a common name anyway. I took the trouble to familiarise myself with Latin names, so, perhaps rather churlishly, I just tell people to buckle down and learn them too. However, I have been slightly thwarted in my evangelism.

The Wildlife and Countryside Act 1981 introduced the requirement that all UK species in need of protection have a common English name.* While nearly all British plants and large animals were possessed of one, most fungi were not, so the British Mycological Society set about creating 'common' names for all but the most obscure.[2] Some of them put my own inventions to shame. We now have the dung cannon (*Pilobolus crystallinus*), the cabbage parachute (*Micromphale brassicolens*), the crystal brain (*Exidia nucleata*), the mousepee pinkgill (*Entoloma incanum*) and, my personal favourite, the midnight disco (*Pachyella violaceo-nigra*). While I don't mind telling someone that the little mushroom in their hand is called *Tubaria furfuracea*, I do feel embarrassed when informing them that they have a 'scurfy twiglet'. There is nothing particularly wrong with these names, but they lack the weight and authority that comes with long usage. Also, I don't think they really help: if people are having difficulty with names, the last thing they need is a whole new set of them. In my opinion, it is simply not possible to make up common names and expect them to become a useful currency.

* An imposition that the bryologist (moss specialist) A. J. E. Smith described as absurd, noting that such names are more difficult to remember and often more cumbersome.

The glorious confusion of common names is just part of human culture and something to be treasured, not dismissed. But confusion, glorious or otherwise, is of no use in science or in the activities informed by science. A scientific paper on the social habits of some Peruvian lizard is useless if it does not tell you *which* Peruvian lizard. It is of no help to give the local name, because that means nothing to anyone except the locals.

Every bit as important is the fact that most organisms (by a very long way) do not have a common name. Camel, cat and capybara are familiar enough, but there are millions of insects and other invertebrates, for example, which are far too unimportant in everyday life to warrant one. Who, apart from a specialist, needs to call a little bug that lives inside the wing cases of cockroaches *anything*? Most species do not have a common name because most people could not care less if they had one or not. However, it is among the tasks of science to explore and understand the world outside everyday experience, and consistent names for everything it studies are essential. This, ultimately, is what Latin names are for.

Why Latin?

The question must be asked. Would not English* do just as well? Of course, it is mostly a matter of history: Latin was the international language of academic discourse throughout the Renaissance and Enlightenment, when the system of naming was devised. It would be an enormous, pointless and near-impossible endeavour to replace Latin names with English equivalents, and would be a disaster for taxonomy were it ever attempted. The problem with

---

* Esperanto has also been suggested.

English names, or names in any language apart from a 'dead' one like Latin, is that they can easily become unstable, suffering what might be called 'nomenclatural drift'. All living languages evolve, and evolution is not something to be desired in biological nomenclature, where stability is the most important requirement. *Drosera rotundifolia*, for example, if officially renamed 'round-leaved sundew' (in fact this is the common name), could so easily over time become 'oval-leaved sundew' (its leaves are slightly more oval than round). There is also the inevitable problem of officially approved (English) species names becoming confused with various common names for the same organism. It is, therefore, partly in its forbidding nature, the very thing that alienates the layman, that Latin has proved so reliable a medium.

There is another extremely useful property of Latin names that would be difficult to accommodate in English. They are binomials, consisting of two words, the first of which tells us the genus to which the species belongs (much more on this later and truly something to look forward to). Although similar constructions are familiar in English – 'black duck' and 'wood sorrel', for example, where 'duck' and 'sorrel' can be viewed roughly as genera – endless problems would occur in any attempt to formalise them in a taxonomy. Wood sorrel and common sorrel are only very distantly related, while the black duck is in the same genus as the mallard, the garganey, the teal and the wigeon but in a different genus to the tufted duck. These common names are useful in their way, but they fail to tell us anything about the organisms' relationships with *other species*, and most would have to be dispensed with in any officially approved list. We would perhaps have the 'teal duck', the 'wigeon duck' and the 'tufted scaup' (to borrow a common name from its close relative), but I have no idea what we could do with the sorrels.

Finally, although English names are available for familiar types of organisms, science deals with the unfamiliar, for which English words do not exist (certainly not at the level of differentiation required by science). Names for slime moulds, bacteria, several million insects and very much more would need to be invented. What, for example, would one call the protozoan reindeer parasite *Besnoitia tarandi*, bearing in mind that there are several hundred genera of such organisms and several thousand species?

To be strictly honest, while Latin names are immune to the aforementioned Chinese whispers that official English names might suffer, they are not quite as stable and unambiguous as one would like. This is important, because apart from providing scientists with consistent names for the objects of their study, Latin names are often items of crucial practicality. In conservation legislation, for example, ambiguities can be disastrous. You cannot protect something that you cannot accurately name, and stories abound of the wrong fish, wrong bird or wrong lichen being protected while the endangered species concerned was left to its fate.[3] A marine snail, for example, which was used to make cosmetics, was found to be exactly the same species as one that had been listed as endangered – conserved under one name but commercially exploited under another. I will explain why Latin names sometimes change later, but it is important to note that they do not do so on a whim or to keep one step ahead of the police – there is always a good, scientific reason. The ultimate aim of taxonomists is to put things in order and provide stable names, but these problems will never entirely disappear. Taxonomy will always be a work in progress.

Does the general public need to learn the Latin names of fungi, plants and animals? Not usually. Anyone asking their grocer for some *Apium graveolens* (celery) and a couple of pounds of *Solanum tuberosum* (potatoes) would be marked down immediately as a troublemaker.*

If your interest is natural history, it is a different matter. Your enjoyment and understanding will be greatly enhanced by learning the names of the organisms that you discover on your travels. And it does make you sound considerably more authoritative when speaking about them to others. Ignoring the comment by the American writer Brander Matthews that 'a gentleman need not know Latin, but he should at least have forgotten it', you do not need to to be familiar with the language in order to understand Latin names. They start to permeate the mind as you learn them and learn how their authors created them. Latin species names and descriptions (yes, there are Latin descriptions, too, see p. 94) are often fairly easy to understand if you are familiar with the technical descriptive words found in floras and the like and you possess a couple of good reference books. Even plain English can be a useful guide. *Noctiflora*, for example, is straightforward because we are familiar with the words 'nocturnal' and 'floral', hence 'night-flowering'. Similarly, it takes little effort to work out what *semisanguineus* might mean ('half blood-coloured'). It can be an interesting diversion to try to deconstruct and analyse

---

* There is one day-to-day exception to this. Go to the garden centre and ask for a *Camellia japonica* or a packet of *Antirrhinum majus* and the smiling assistant will merely ask which colours you had in mind. Most of the plants that fill glossy catalogues come from overseas and arrive stripped of their native common names. With no English equivalents, the first names they acquired were those given to them by botanists, and those are the names that stuck.

etymologies, although many excellent books have been published that succeed in taking all the fun out of the exercise.

Here are a couple of straightforward examples to show how names can be decoded.

Boletus impolitus is a mushroom with a slightly puzzling name, and I was seriously caught out many years ago when asked to explain its derivation at a lecture I was giving. (It is always embarrassing not to know the answer to a question when you and everyone else think you should.) *Boletus* is just a Latin word for a mushroom, but *impolitus*? There did not seem to be anything rude about it – more than can be said for a large number of fungi (see p. 53) – so I was stumped and suitably shamefaced. The word, it transpired, comes from the Latin *politus*, which means 'polished', so *impolitus* is 'unpolished'. And, indeed, the mushroom has a suede-like surface to the cap. In fact, I was not too far off in considering the idea of rudeness, because 'polite' derives from *politus* as well – if you are polite, you are polished.

Somateria mollissima is another highly descriptive name. *Soma* means 'body' (in Greek; many 'Latin' words were taken from Greek and then Latinised) and *erion* is 'wool' (also Greek). *Mollis* is Latin for 'soft', and the word ending *issima* means 'very'. So we have 'thing with very soft body wool' (with the word 'thing' implied – a convention I will use throughout this book). It all adds up to a fitting name for the much put-upon common eider, from which all those bed coverings have been made.

Many Latin names are as descriptive as these two, but some are particularly evocative. What better name for the venomous puff adder could there be than *Bitis arietans* ('striking, biting thing') – other than perhaps *Lachesis muta*, which belongs to the

South American bushmaster and means 'silent fate'? The Canadian porcupine goes by the perfect name of *Erethizon dorsatum* – 'having an irritating back'. However, as this book will reveal at length, many scientific names are as useless as *Troglodytes troglodytes troglodytes* ('hole burrower hole burrower hole burrower', the burdensome name of the northern wren) and *Puffinus puffinus*, which is, of course, the name of the Manx shearwater.*

As the unfortunate wren would testify, organisms are not in a position to choose their own names – *edulis* (edible) and *deliciosus* would not be anyone's first choice. Neither does any living thing come with a label attached. Names are the work of men, and many seem designed to amuse or impress as much as to inform. So it is, by accident or design, we have the brown noddy *Anous stolidus* ('mindless fool'), the beetle *Anelipsistus americanus* ('helpless American') and the stump puffball *Lycoperdon pyriforme* ('pear-shaped thing as flatulent as a wolf'). Others are clear Latinised absurdities, such as *Spongiforma squarepantsii*, a species of fungus named after a cartoon character – in fact the comparison describes it remarkably well – and the euphonious *Tamoya ohboya*, a box jellyfish named in 2011 by school teacher and marine biology specialist Lisa Peck, who won a competition to name the species. 'Oh boy,' she pointed out, is what you might say on being stung by one.

The following pages are packed with absurd, unpronounceable, self-seeking, vulgar or plain wrong names. But there

---

* 'Puffin' is, in fact, the old name for the Manx shearwater, only later being used for the puffin as we know it today. Puffins are in the genus *Fratercula*, meaning 'little brother', a charming name derived from the bird's fancied similarity to a monk.

is no need to look for excess in Latin names to be impressed by them. Some are simply beautiful: *Oenanthe oenanthe* (wheatear), for example, which derives from the Greek *oinanthē*, 'first shoots of the vine', indicating the time of the year in which the bird appears. The North American wood duck was named by Linnaeus (the most romantic naturalist of them all) *Aix sponsa* ('betrothed water bird'), because it appeared to be dressed for its wedding.

As I have noted, the most obvious property of Latin names (apart, of course, from the fact that they are in Latin) is that they are binomials (see above),* a convention established by Linnaeus in the middle of the eighteenth century. The first word is the 'generic name', which indicates the genus; the second is called the 'specific epithet', denoting which species *within* that genus. So, for example, the genus *Quercus* (oak) contains around six hundred species, each distinguished by its specific epithet: *Quercus robur*, *Quercus alba* and so on. The binomial nature of Latin names is similar to that of personal names, because they convey information about genealogy. So 'John Wright' is not just a name; it tells us that someone called John was born into the family Wright. Similarly, the crested myna bird, *Acridotheres cristatellus*, is not merely a bird that hunts locusts and has a little tuft on its head (*akridos* meaning 'locust', *therao* 'hunt', *cristata* 'crest' and *ellus* 'little'). It is also a member of the genus *Acridotheres*, together with another dozen or so other species.

The comparison with surnames can be taken no further, for the generic name can itself be nested within a higher-level group.

---

* They can sometimes be trinomials (consisting of three words), representing a subspecies.

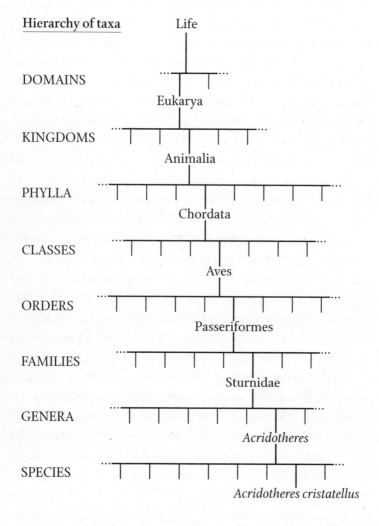

The fundamental ranks in taxonomy, following the lineage of *Acridotheres cristatellus*. The short vertical lines indicate other members of the class, order etc. Intermediate ranks such as super order, infraclass and cohort are not shown.

Acridotheres is a member of the 'family' Sturnidae, which contains other genera that all encompass more distant cousins of *A. cristatellus.** Sturnidae is in turn a member of the 'order' Passeriformes, which is a member of the 'class' Aves, which is a member of the 'phylum' Chordata, which is a member of the 'kingdom' Animalia, which is a member of the 'domain' Eukarya. Every species belongs to one genus, one family, one order and so on. The crested myna bird, by its Latin name, is given its place in the evolutionary tree of life.

Important to this discussion is the matter of a 'taxon'. The word comes from 'taxonomy', the whole process of identifying, naming, describing and classifying organisms, in turn deriving from the Greek words *taxis* ('to arrange' or 'to classify') and *nomos* ('law'). A taxon is any group at any specified level within the hierarchy. So the kingdom Plantae, the order Apiales, the genus *Angelica* and so on are all taxa, higher taxa being made up of all their subordinate taxa.† However it is worth noting that species, too, are considered to be taxa, but here the taxon is made up entirely of individual organisms.

The primary subject of this book is the scientific names of organisms, but since these embody part of their own hierarchy, it is impossible to divorce them from it – it would be like writing a family history with everyone's surname and familial relationships omitted. Few naturalists of the past have written their floras or faunas without at least a nod in the direction of classification. It is natural for people

---

* The genus name is conventionally abbreviated to a single capital letter in subsequent mentions.

† Unnamed groups, for example between order and family, can also be taxa, provided that they form a 'natural' group, i.e. a group that can be cut from the tree of life with a single application of the secateurs.

to like order and to see it in the world around them, and such order is expressed even in everyday speech. We speak of pied wagtails, grey wagtails and yellow wagtails in the full understanding that they are related in some way because they are all wagtails (although bramblings and crossbills do not wear their familial relationships so boldly, leaving us to guess that they are both finches).

The search for order in the natural world began in Greece more than two thousand years ago. Until Darwin, writers of botanicals and the like were happy to use any order, making do with artificial classifications based on matters such as aquatic/terrestrial, flower shape, medicinal properties or even alphabetical order, just to make some sense of their lists and, at least, to make it easy to look things up. Nevertheless, from classical times until the nineteenth century, just about everyone had one particular idea what a 'natural' order – one that reflected how things were truly ordered – would look like, and tried to impose it on the living world. The idea of a natural order comes from the trivial observation that there are many very different living things and they seem to come in groups, the members of which bear certain similarities, and that there is a certain amount of gradation between groups (for example, newts look like a halfway house between fish and lizards). They appear to be organised. But how and, indeed, why?

The 'why' was easy: that was the way in which God had ordered his world. Unfortunately, the order that most pre-Darwinian naturalists settled on was completely wrong and many of the taxonomic errors of botanists and zoologists through the ages can be laid at its door. It was the *scala naturae*, literally the 'natural ladder', also known as 'the great chain of being'. Originating with Plato and Aristotle, it predates Christianity but fitted in very nicely with Christian thought and the social order that reigned when the Church was ascendant and western naturalists busy with their taxonomies. Simply put, it suggests that everything is arranged in a long line, with God at one end and rocks at the other. It is a highly judgemental system: gold is placed above silver, birds of prey are above birds that eat seeds, Man is situated uncomfortably somewhere between the animals and the more everyday sort of angel, and so on. All very neat and handy for keeping things (and people) in their place, but as an attempt to show the actual arrangement of the living world, it produces only nonsense. The true arrangement looks like a (very large) branching tree with most extant organisms represented only by the tips and branching joints of a few twigs.

The *scala naturae* has had a particularly mortifying effect on botanical taxonomy because botanists long considered size to be an important indicator of an organism's position on the *scala naturae*, leading them to place trees above shrubs and shrubs above herbs. However, the careful work of naturalists from the sixteenth to the early nineteenth century (see p. 232) gradually built a structure that began to look very much like the tree of life as we now understand it, despite their preoccupation with the *scala naturae*. When Darwin came up with the bright idea that everything was related to everything else because species evolved from species in an ever-expanding tree, the natural order at last became clear.

The fundamentals of the nested hierarchy that most modern taxonomists use to arrange the living world were largely devised by Linnaeus, albeit in remarkable ignorance of the true natural order. This, together with his convention of naming species using binomials and the laying down of rules for how those names should be created and employed, make up the 'Linnaean system', to which science owes such a debt.

Sorex araneus

CHAPTER II

The NAMES THERE ARE

THE COMMON SHREW, OUR EPONYMOUSE HERO, IS FAMILIAR TO MOST people as something the cat brought in. Cats 'donate' shrews to their owners, rather than eating them, because, by all accounts, they taste awful, even to a cat. Quite where the shrew fits into the natural world is something most people neither know nor care about, but if asked they would probably say that it was a type of mouse, a rodent at least. I have a dried one on my desk, and alongside it a mole, which a zoologist friend kindly freeze-dried for me (it's not what you know, it's who you know). Apart from the fact that one is twenty times the size of the other and has spades for hands, the similarity is remarkable, as indeed it should be, for these are closely related species. They belong to the order Soricomorpha – the 'shrew-shaped'. Mice are in a different order altogether, the Rodentia, and closer to porcupines than shrews. Such fine taxonomising is of little interest if something small, fast and furry is discovered running round the kitchen, however. For most people it is a type of mouse, and onto the chair they leap.

While a distinction has been made since early times, confusion has reigned. The Greek *sorcio* and its Latin cognate *sorex* for shrew, and the Latin *mus* for mouse, have been used inter-changeably and often incorrectly. The French call the shrew *musaraigne* and the mouse *souris*, so the two uses are quite the wrong way round. *Sorcio*, incidentally, may be onomatopoeic, referring to the soft rustling noise the shrew makes – 'susurrate' being a related word.

When Linnaeus named this creature he drew on two existing Latin nouns: *sorex* ('shrew') and *araneus* ('spider'). He did not explicitly call it *Sorex araneus* until the tenth edition of his most important work, *Systema Naturae*, and even then, as was his convention (see p. 221), the epithet was merely provided in the margin. Prior to this, he listed it under the genus *Sorex* and referenced the name given to it by his taxonomic predecessor John Ray (see p. 175): *Mus araneus* ('spider mouse'). Linnaeus knew that there was a big difference between mice and shrews, so he dispensed with *mus* and gave the shrew its proper Latin name.

That explains where *Sorex* came from, but what is the origin of *araneus*? It may sound extraordinary, but until recently the shrew had a most fearsome reputation. The creature's bite was likened to that of a spider – *araneus* in Latin. Both Aristotle and Pliny wrote of its venomous nature, and this belief continued down the centuries, gaining momentum as time went by. The general feeling was summed up neatly by the Reverend Topsell in his seventeenth-century *History of Four-footed Beasts and Serpents*: 'It is a ravening beast, feigning itself gentle and tame, but, being touched, it biteth deep, and poisoneth deadly. It beareth a cruel minde, desiring to hurt anything, neither is there any creature that it loveth, or it loveth him, because it is feared of all.'

Not that an actual bite was considered necessary for the shrew to do its evil work. Elyot in 1538 wrote: '*Mus Araneus*, a kynde of myse called a shrew, whyche yf it goo ouer a beastes backe, he shall be lame in the chyne'. 'Chyne' here means 'spine', so it

[18]

would have been a calamity if it were true, which, of course, it was not. Horses were considered particularly vulnerable, but it was not just dumb beasts that were at risk.

Although there is much to be feared in the modern world, one threat at least is no longer a burden, and anyone attending the doctor's surgery complaining of being 'shrewstruck' would not be received sympathetically. This fictitious condition was the result of having a shrew 'goo ouer' some part of your body, causing pain and even paralysis. Fortunately such imaginary ailments respond well to imaginary remedies. Gilbert White in *The Natural History of Selborne* reported the destruction by a pious vicar of a much-venerated ash tree. The tree – a 'shrew ash' – was relied upon as a cure by the village people, who pleaded in vain for its survival. To make such a tree, a hole was drilled into the trunk, then a live (and very unlucky) shrew was placed into the hole and incarcerated there with a wooden plug, to the accompaniment of appropriately dramatic incantations (sadly lost to history). The branches were then available to be 'applied' as a cure, although precisely what this entailed is not recorded. With so many fine details forgotten, should you ever imagine that you have been 'shrewstruck', you will be in no position to imagine that you are cured.

Any plant or animal possessed of such powers as that of the shrew tends to be a two-edged sword, as likely to cure an illness as to cause one. One remedy I particularly like is that of cutting off the tail of a live shrew,* burning it (the tail) and applying the

---

* I am reminded at this point of yet another French name for the shrew: *mus caecus*, or 'blind mouse' (the shrew is indeed practically blind, relying on echolocation to find its way around). I humbly offer this as a reasonable explanation for the popular and otherwise incomprehensible nursery rhyme.

resultant ash 'upon the sore of any man, which came by the bite of a greedy and ravenous dog'.[4]

As science progressed apace in the nineteenth century, the idea of the shrew being venomous started to seem ridiculous, and any actual effect was ascribed to a bacterial infection of the wound. However, occasional reports continued to appear of people being poisoned by certain species of shrew in various parts of the world. In 1889, a shrew from the New World – the northern short-tailed shrew, *Blarina brevicauda* ('short-tailed one from Blair') – became a major suspect when the naturalist C. J. Maynard suffered an almost immediate swelling of the arm after being bitten.[5]

Now, the front two teeth on the lower jaw of all shrews are vicious-looking instruments, pointing horizontally forward from the jawbone. They are red-tipped, as though with blood (in fact it is iron oxide, which hardens the enamel), and as sharp as needles, seeming designed neither to bite nor to chew, but to impale. In some, the top edges of these teeth have little grooves, which, it transpires, allow the animal's saliva, laced with a toxin, to enter the victim's body, causing paralysis or even death. Mice injected with a substantial amount of the saliva are usually dead within minutes. It is now known that the northern short-tailed shrew is one so venomously endowed, and it is probable that the mercifully extinct giant shrew, *Dolinasorex glyphodon*, was similarly talented.

The tiny *Sorex araneus* is unlikely to be a pressing cause for concern, but a related and larger native animal, the water shrew (*Neomys fodiens*, 'swimming mouse that digs'), most certainly is. The two were long considered to be the same species, which is surely the reason for so much vexation over the centuries.

The story of the shrew is a perfect example of the richness that can lie behind a name. I take enormous delight in finding out, or better still, working out, the etymologies of scientific names. A warm feeling came over me, for example, when I discovered that the Latin name for Oxford ragwort, *Senecio squalidus*, means 'dirty old man'. (For the purpose of gender equality, I also offer the humble primrose – *Primula vulgaris*: 'first common girl'.) Who would not marvel to hear that the dinosaur *Qantassaurus* is named after Qantas Airlines, that there is a butterfly called *Charis matic* (and another, no doubt less impressive species, called *Inglorious mediocris*) and that *Han solo* is a species of trilobite?

With a bit of Latinisation and an affix or two, just about any word can be turned into a generic or specific name. Nouns and adjectives, because of the job they do, are the most common, and although the words that make up a binomial can come from anywhere, a few types of name dominate. In a survey of all nine hundred or so known specific epithets of the genus *Aloe*, Estrela Figueiredo and Gideon F. Smith[6] found that morphological words (describing shape or spininess of leaves or colour of flowers, and so on) were the most frequently used, with 353 instances. There were 278 species of aloe named after people, 179 for their geographic location, 37 for habitat, 34 for their relationship with other species, 16 for their beauty, six after a vernacular name and twelve 'others'. This survey is likely to be typical of botanical epithets at least, revealing the 'big three' inspirations for specific names: morphology, eponymy and geography.

Nounal names

The genus name *Aloe* was established by Linnaeus from the Greek name for the plant. As with *Sorex*, the first stop for a nomenclaturist looking for a genus name is often the direct Latin or Greek equivalent. If you wish to name a genus of dog then *Canis* seems like a good idea, if a pig it would be *Sus*, for a fig you would likely choose *Ficus*. The genera *Sciurus* (squirrel), *Bufo* (toad), *Strix* (owl), *Felis* (cat), *Agaricus* (mushroom) and *Linum* (flax) are derived directly from the Latin. But despite often being described as 'Latin names', many are taken from the Greek and then Latinised. Single noun genera taken straight from Greek are relatively rare however, as most of them were already Latinised by the Romans, notably Pliny. *Arum*, *Clematis* and *Geranium* are among the few familiar genus names that did not first pass through Rome.

Though slightly obscure, the name *Sorex araneus* at least tells us something about the animal: it is a shrew and there is something 'spidery' about it. However, taxonomists have often delighted in over-egging the pudding by effectively using a common name twice – once in Latin and once in Greek. *Bos taurus*, for example, means 'ox ox', and *Ficus carica* means 'fig fig'. Similar oddities are *Salmo salar* ('salmon salmon') and *Sus scrofa* ('pig pig'). With only two words to play with, it seems like a waste. These wilfully created names are not to be confused with the more blatant tautologies found in zoology, such as *Gorilla gorilla*, which come about because of the vagaries of the rules of nomenclature (see p. 107). Even more absurd are the names that look like zoos. *Giraffa camelopardalis*, literally 'giraffe camel leopard', is a name that smacks of not being able to make up its mind. In fact, it is of the same construction as *Bos taurus*, where the same meaning in different languages is used: *Giraffa* derives from the Arabic for 'giraffe', *zarāfah* translates as 'one who walks swiftly' and *camelopardalis* is a Greek construction for 'giraffe', meaning, more or less, 'camel wearing a leopard coat'. If you think the giraffe's name silly then ponder *Boselaphus tragocamelus* ('ox-deer goat-camel'). Having seen a picture of one I can tell you that it describes the poor creature perfectly.

A large number of Latin and Greek words refer to more or less the same thing. There are about eight classical names for 'ape', and Roland Wilbur Brown in his compendious *Composition of Scientific Words* (1954) listed more than fifty words for which the only translation he gave was 'a plant', although whether this owes to classical plenitude or simply not knowing which plant a word referred to is uncertain. Similarly the guillemot genus *Uria* derives from the Greek *ouria*, which appears to mean any old water-loving bird.

Some single generic nouns refer to organisms that clearly are *not* the species being named: *Delphinium* (Greek for 'dolphin') and *Melissa* (Greek for 'bee'), both genus names for plants – the former from a fancied resemblance of the flower to a dolphin and the latter because of its presumed source of nectar. *Tragus* is a genus of grass, but the word derives from the Greek *tragos* for 'he-goat'. Evidently the author of this genus, Dioscorides (see p. 160), thought that the grasses' bristly spikelets looked like a goat's beard.

Nounal names are not restricted to plants or animals: shapes, colours, body parts, household objects, hats and a thousand other nouns have found new and gainful employment as names for organisms. From Australia comes *Petaurus australis* (the yellow-bellied glider), derived from *petaurum*, the Latin word for 'spring-board' (this possum leaps about a great deal). Other examples include *Gladiolus* ('little sword'), *Iris* ('rainbow') and *Helix* ('spiral'), an irresistible name for a genus of snail. Linnaeus and many of his successors were not terribly fond of 'barbaric' names – those derived from non-classical sources – but he did use them himself on occasions. As well as the Arabic *Musa* for the banana (see p. 194), he also accepted the Japanese *Ginkgo*. Now that taxonomists have overcome their qualms about barbarity, they have an endless resource of vernacular names from many languages to draw upon. Local common names can relate a genus to its geographical and cultural home (although this meaning can be lost if the genus is found subsequently to have a wide distribution). *Tayassu pecari* (white-lipped peccary), for example, comes from two South American names for the peccary; *Iguana* is from the Spanish-Arawak (West Indian) name for lizard. The beautiful and endangered indri (*Indri indri*), a species of lemur, derives its common, its generic and its specific name from the Malagasy word for the animal, *endrina*, although the long-held view that it comes from the Malagasy word for 'there it is!' is much more appealing.

Mythological names

The tradition of naming plants and animals after people, real or imagined, is nearly as old as taxonomy itself. Linnaeus had a particular liking for mythological names, many of which he took from classical writers, such as Theophrastus (see p. 154) and Dioscorides (see p. 160). He and his successors ransacked classical sources for gods, goddesses and other mythical beings to fill the need for new genera and, often, species.* Like a closing-down sale,

---

* In English the very words for plants and animals – flora and fauna – are the names of Roman gods.

all the best stuff went first, leaving later taxonomists to make do with second cousins and third daughters of lesser-known gods. Linnaeus was, of course, at the head of the queue and appears to have gone straight for the big one: Zeus. The most familiar species in the genus *Zeus* is *Zeus faber* (John Dory), a nearly three-hundred-year-old specimen of which languishes in the Linnaean collection in London and is starting to show its age. In fact, Linnaeus simply applied the name provided by Pliny, although why an admittedly attractive and tasty fish should be named after the 'Father of the Gods and men' is unclear.

Many of Linnaeus's mythological names were his own, and occasionally he supplied an etymology. He named the rather creepy nocturnal *Lemur tardigrades* (which is not a lemur, but a loris*) after the 'lemures', which were creepy, nocturnal and spectral creatures of Roman mythology. He also used mythology as a handy source of ready-made names, which did not necessarily need to match the species they were applied to. He placed all the butterflies in the genus *Papilio* (Latin for 'butterfly'), then, like Cecil B. DeMille, set about employing a cast of thousands (nearly two hundred, anyway) to supply his specific epithets. Many of these were mythological, and he even grouped similar species into 'families' according to their classical contexts. His *Equites Trojani* (Trojan army), for example, contains species such as *Papilio hector*, *P. paris* and *P. helenus*, and *Equites Achivi* (Achaean army) has *P. ulysses*, *P. agamemnon* and *P. menelaus*.†

---

* Other species that *were* lemurs ended up in the same genus and kept the name for themselves.

† Specific epithets can be used many times. A generic name can also be used more than once but the genera involved must be very clearly distinct, usually belonging to different kingdoms.

The zoological space for Zeus was filled, but it seems that no botanist or mycologist spotted that *Zeus* was still available to them. Then, in 1987, the British Mycological Society organised a brief expedition to Mount Olympus in Greece. An old friend of mine, David Minter, a classical scholar turned mycologist, went along and spotted a small, cup-shaped fungus embedded in the trunks of some pine trees. It was a new species, confined to a tiny geographical area. It was also a new genus. The temptation afforded by the opportunity of naming a species and genus at the same time was too much for David and he named it *Zeus olympius*.[7] It was a productive trip, for growing on the decaying remains of specimens of *Z. olympius* was a minute, flask-shaped fungus, a new species of *Nectria* – a genus familiar to gardeners in the troublesome form of *Nectria cinnabarina*, or coral-spot fungus – which David and his colleagues were to call *N. ganymede*. The etymology section of the paper explains the new epithet: 'Ganymede, "a beautiful youth of Phrygia who was taken up to heaven by Zeus to become cup-bearer to the gods and to sit on Zeus' lap" (John Lemprière, 1788).'

Names based on mythological characters are found often in zoology and occasionally in mycology. However, every other plant seems to bear one, and Peter Bernhardt filled an entire book with them in his *Gods and Goddesses in the Garden*. The gardener will be familiar with many of these spiritual plants, but not always the menagerie of mythologicals for whom they are named. Just a small sampling of the 'A's gives us Linnaeus's bog-rosemary (*Andromeda*), marshmallows (*Althaea*, daughter of Thestius and sister of Leda), dragon's mouth orchid (*Arethusa*, a nymph) and wormwood (*Artemisia*, from Artemis, who was nearly the nemesis of St Paul during an ill-advised weekend break in Ephesus).

Astartea, Helianthus, Hyacinthus and *Orchis* are a few more genera with mythical associations. Orchis, who gave his name to a genus of orchids that have tubers shaped like a pair of testicles, was the son of the gifted Priapus.*

Most classical myths are soap operas for intellectuals, as demonstrated by this example related by Bernhardt. The Titan king Cronus was unfaithful to his wife (and sister) Rhea, with his niece Philyra. Rhea caught him at it and, in a few action-packed moments, he turned himself into a stallion, ejaculated in haste and ran away. Philyra gave birth to a boy who was half horse; he became the first centaur, Chiron. Obviously upset at this turn of events, Philyra transformed herself into the first linden tree. Chiron married a nymph and had a daughter, Ocyrrhoe, and Rhea (whether infuriated by the treachery of her husband/brother or just out of natural bad temper, I don't know) turned her acolyte, Cyparrisus, into a cypress tree after he killed a stag. (The story is better when Bernhardt tells it.)

Here are a few of the names that come from this tale: *Amorphophallus titanium* (a plant with what Bernhardt describes as a 'titanic, misshapened phallus'); *Rhea* (a genus of flightless bird and the specific epithet of a frog); *Phillyrea* (the linden tree already had a name, *Tilia*, so Linnaeus used it as the generic name for some evergreen shrubs); *Cupressus* (the cypresses); *Chironia* and *Centaurium* (related genera placed by Linnaeus in the family Gentianaceae, after King Gentius – a real person who used gentians as a medicine). There are more, but I am sure the point is made.

Incidentally, the badly behaved Cronus was the son of Gaia, the Greek goddess of the Earth. She makes a moderately overt

* Priapus pops up often in biological terminology.

appearance in such names as *Epigaea* – a shrub that is 'on top of the world' – and a much more substantial one with her related name of *Ge*, meaning 'Earth' or, more prosaically, 'the ground'. *Geodorum* (orchids, literally 'Earth's gift'), *Geocaryum* (in the carrot family: 'Earth's nuts'), *Geocharis* (a beetle genus, charmingly 'Earth's charms') and the fungus genus *Geastrum* (earth-stars) are few of the many.

The animal kingdom (apart from butterflies) is, perhaps, slightly less well endowed with mythological names than the plants. One example is the *Priapulida* (from the aforementioned Priapus, the permanently enthusiastic Greek god of fertility), an entire class of marine worms that have a spiny proboscis shaped like a penis, although, frankly, the name suits the entire animal. More respectably, the Titan Oceanos is remembered in *Oceanodroma* (loosely translated as 'running the course of the ocean'), a genus of storm petrels. Classical mythology has cornered the market in providing names for species, but gods from other cultures make an occasional appearance. The colourful if rather ugly thorny dragon, *Moloch horridus*, was named for Moloch, the Ammonite god with notoriously bad habits. *Horridus* is Latin for 'spiny'.

Few animal names are more evocative than *Furia infernalis*. One imagines a flying dragon, or perhaps a sabre-toothed tiger, only bigger. On one of his early expeditions, Linnaeus was bitten on the arm by an unidentified creature. The bite was painful, but he thought little more about it until his arm swelled up. He became seriously ill for some time until a surgeon called Dr Schnell incised the arm from armpit to elbow and he recovered. A few years later, Linnaeus settled on a tiny 'worm' that had been described by his pupil Daniel Solander as the cause of all his suffering, and proceeded to seek redress against his tormentor. He gave it the rather splendid name *Furia infernalis* (the fury of Hell – the Furies were mythical creatures of vengeance), describing it as threadlike with ciliate, appressed spines on both sides of its body. Worst of all, he wrote, it fell from the air, penetrated the bodies of animals and caused excruciating pain.

The creature, evidently thrilled with its new name and no doubt emboldened by all the publicity, proceeded to make a thorough nuisance of itself for the best part of a hundred years. A celebrated traveller from England, one Dr Clarke, reported that he, too, temporarily lost the use of his arm due to the attentions of a *Furia infernalis*, with only the Lapp remedy of a curd poultice saving the day. Cattle also suffered, and in 1823 a five-thousand-strong herd of Lapland reindeer was slaughtered by pestilential hordes of *Furia*. A young girl, having been stung on the finger by a *Furia*, survived only because of the quick thinking of her master, who cut off the afflicted finger on the spot. A ban on fur imports was introduced by Finland to prevent the spread of this calamitous worm, with some success because only Russian and Swedish Lapland were affected.

Subsequent to Linnaeus's establishment of *Furia*, other naturalists produced dissertations on the creature, establishing its modus operandi (it crawls to the top of a reed and allows the wind to blow it onto the skin of mammals), several more cases of human infection were recorded and a better description than that supplied by Linnaeus – 'it is the thickness of a human hair, grey with black extremities' – was made. Even at the time of its first publication, however, not everyone accepted the existence of the animal. They were supported in their cynicism, it must be said, by a total lack of any specimens. The Academy of Sciences in Stockholm offered a reward for one, but none was forthcoming, and a Mr Retzius of Stockholm undertook an extensive search for the elusive creature without success.

Furia infernalis does not exist and has never existed, but, owing to the rules of zoological nomenclature (see p. 107), its name will live on for all time. Linnaeus (who, to his credit, came to doubt the existence of the species later in life) named it in the tenth edition of *Systema Naturae*, the edition that is accepted as the starting point for the naming of animals. *F. infernalis* is counted as a valid name* and cannot be used for anything else, but it is not an accepted name, having never had a species to which it could belong. Linnaeus was, no doubt, bitten by a horsefly.[8]

Fictional names

Mythological names, venerable as they are, enjoy a certain credibility despite often being arbitrary and always fictitious. However, those who may be content with *Dionysia* and *Narcissus* may not be quite so enthusiastic about *Polemistus chewbacca* or *Otocinclus batmani* (respectively a wasp and a South American catfish). With fictitious names we see nomenclaturists having fun – *Oedipina complex* and *Oedipus rex* were no doubt irresistible to the zoologist Dunn when naming a couple of salamanders in the 1920s – and they are very popular, at least they are with nomenclaturists. Three fossil ginkgo species named in a 2001 paper by three female authors[9]

---

* Technically it is an 'available' name, because it was validly published. There are big differences between being published, being nomenclaturally correct and being the true, accepted name of a species. Unfortunately the codes of nomenclature do not agree on what terms should be used. For example, the accepted name for something is 'correct' in botany and 'valid' in zoology.

were *Ginkgoites weatherwaxiae*, *G. nannyoggiae* and *G. garlickianus*, commemorating three witch characters created by Terry Pratchett (Granny Weatherwax, Nanny Ogg and Magrat Garlick). Other species named after Discworld characters include fifteen braconid wasps of the genus *Aleiodes*, such as *A. rincewindi* and *A. vetinarii*.

Star Wars characters are a favourite source of inspiration. The trapdoor spider species *Aptostichus sarlacc* is named after a monster called the Sarlacc[10] – both lurk underground to catch their prey – and *Yoda purpurata* is a genus of deep-sea acorn worms. The large lips on the sides of its head are reminiscent of Yoda's ears, although as the epithet attests, they are purple, not green. Around fifty genera and specific epithets are based on Tolkien characters, from the fossil tardigrade genus *Beorn* to the extinct mammal genus *Tinuviel*. The Harry Potter stories inevitably receive nomenclatural attention with such names as *Aname aragog* (another trapdoor spider) and a remarkably dragon-like, but disappointingly herbivorous, dinosaur called *Dracorex hogwartsia* (dragon king of Hogwarts).

Perhaps you are starting to see a pattern. 'Serious' fiction has little attraction for taxonomists, no doubt because of the fantastical nature of the organisms they study rather than an unhealthy obsession with fantasy fiction. Even eighteenth- and nineteenth-century literary references tend to be from the more imaginative authors, such as Swift, Stoker, Kipling and Melville. The only straightforward literary author who gets much of a look in is Nabokov[*]. His masterpiece *Lolita* is commemorated at least four times in species such as *Humbert humberti* (a wasp) and *Madeleinea lolita* (a butterfly). Somehow this is more worrying than any number of space aliens, dragons or elves.

---

[*] Nabokov was himself an accomplished taxonomist – of butterflies.

Honorary names

Eponyms in taxonomy can be real as well as fictional, of course, and the practice of naming species and genera for real people has a venerable pedigree. Linnaeus made full use of such names, although sometimes the honour they bestowed was less than welcome. He instructed that they should be reserved for notable naturalists (such as Linnaeus, for example), and this tradition predominates even now.

Linnaeus was not a great traveller, so he needed willing volunteers (or victims, as things all too frequently turned out) to explore the world for him and for science and to bring back new specimens. They came to be called his 'apostles'. There were seventeen of them (if one includes Adam Afzelius, who set out after Linnaeus's death), five of whom never returned and two of whom died from their travels and travails shortly after. Most of them produced remarkable results and a wealth of new species, and many were immortalised by Linnaeus.

The first, Christopher Tärnström, did not provide an auspicious start, dying of a tropical fever before he could send back any specimens. His widow blamed Linnaeus for leaving her bereft and her children fatherless, and Linnaeus never again sent a married man. Tärnström is remembered in the genus of tropical plants *Ternstroemia*. Pehr Osbeck was dispatched to China with the primary objective of collecting a tea plant. He returned two years later with a substantial collection of plants, none of which was tea. Despite proving to be incapable of finding tea in China, Osbeck became one of the most famous apostles and was honoured in the East Asian plant genus *Osbeckia*. Less fortunate is Daniel Rolander, who, returning from Surinam, made the journey as far as Germany when his money ran out. He was unable to obtain sufficient funds to proceed further, and when he finally arrived home five months later, he fell out with Linnaeus over his poor treatment and refused to hand over his collection and papers. Linnaeus rewarded this brave adventurer/botanist with a mere epithet in the beetle *Aphanus rolandri*, the translation of which is 'obscure Rolander'.

Open any book with more than a handful of Latin names on the page and one or two will always be found that commemorate a respected ornithologist, botanist or natural historian. For example, *Bowdleria punctata*, the fernbird, is for Dr Richard Bowdler Sharpe, who worked in the bird room at the British Museum at the beginning of the twentieth century. The specific epithet refers to the brown spots on the bird's breast, but it is difficult to believe that it is not an oblique reference to the second part of Bowdler Sharpe's name. This is all the more honourable an eponym because it names a genus.

As a method of reinforcing bonds between colleagues, eponyms are harmless, if not particularly helpful to the non-specialist (unless they are as memorable as the genus of red algae named for a Dr Humm: *Hummbrella*). There is, however, a strong suspicion that many species are named unnecessarily for the sole purpose of honouring a superior. The fir tree genus, *Abies*, is a trial for taxonomists because its component species hybridise easily. One species, accepted by most authorities, is *Abies borisii-regis*, the Bulgarian fir. However, there is a suggestion that it is a mere hybrid of *A. alba* and *A. cephalonica* and thus not worthy of its own name.[11] The tree was named for Tsar Boris III of Bulgaria, having been described during his reign in a clear case of pleasing the biggest boss around. However, Tsar Boris is credited with saving his Jewish subjects from the Nazis, so it would be a pity to take this memorial away from him.

Honouring royalty is a long-standing tradition. The world's largest water lily, *Victoria regia*, with its floating leaves that resemble nine-foot-diameter flan dishes, was named for Queen Victoria by the botanist John Lindley in 1837, the year she became monarch. Such honours continue today. In 2013, Dutch mycologists discovered five new species of *Penicillium*, which were, most unusually, orange. Each of the species received an epithet honouring, in turn, the King, Queen and three princesses from the house of Orange-Nassau.

I am told by friends in the trade, although I am unable to confirm, that Russian, Chinese and certain other naturalists frequently manage to find more species within a defined group than their western counterparts, producing endless eponymic epithets of suspect worth. Opinions about what constitutes a 'good' species vary enormously between cultures, of course, and eastern taxonomists are known to take a narrower view. Perhaps they just feel they have more species in need of a name.

Naming a doubtful species for one's professor is not the worst conceivable sin for a taxonomist. Neither is getting things wrong. The most unforgivable act is to name a species after oneself. Taxonomic onanism is extremely rare, and there are just a handful of examples in the history of nomenclature. The *American Journal of Conchology* could barely contain its indignation in 1865: 'We do not remember ever before to have seen an author name a species after himself, as Mr Bourguignat has done in *Ferussacia Bourguignatiana*'. Evidently having missed a trick, the *Journal of the New York Entomological Society* noted in 1908 that:

'Hochenwarth (1789) is the only man who named a species after himself'. It was an exceptionally furry-looking moth called *Phalaena hochenwarthi*.

The seemingly open-and-shut case of the scarab beetle, '*Cartwrightia cartwrighti* Cartwright 1967' has an endearing explanation. In 1958, the genus *Cartwrightia* was established by Federico Islas and named for the entomologist Oscar Cartwright. In 1967, Cartwright published a paper describing two new species for the genus, one of them being *C. cartwrighti*. But the Cartwright in question was not himself: '*Cartwrightia cartwrighti* is named after my brother Raymond Kenneth Cartwright, who, though not an entomologist, accompanied me and served as guide on many pleasant and productive field trips'.

Self-naming may be beyond the pale, but commemorating a wife or husband is perfectly acceptable. The Figueiredo paper on *Aloe* notes that a little more than one per cent of species were named for wives, although none was named for a husband. A straightforward example of this romantic gesture comes, fittingly, from the French ornithologist Charles Vaurie, who was on honeymoon with his entomologist wife Patricia in the Canary Islands when he discovered a new subspecies of wagtail. He named it *Motacilla cinerea patriciae*.

Naming a species after somebody else's wife (or husband) also has a long tradition, although it is one fraught with danger, so few clear examples exist. The nineteenth-century entomologist Frederick Smith was known to have an affection for the wife of his colleague Carlo Emery, and when he named a species of ant he honoured the lady, not with a direct eponym, but in a flattering reference to her shape. He called the species *Myrmica gracillima* (slender ant), because he thought it to be as slender and graceful as she. This sort of thing seems more dangerous than just using the name, but evidently Carlo Emery was unperturbed.

Fairly late in his life, Linnaeus wrote to Lady Anne Monson, a distinguished amateur botanist and a granddaughter of Charles II. It was regarding a pelargonium-like flower she had collected while in Africa. Linnaeus never met her, which makes the content of his letter all the more extraordinary. 'I have long been trying to smother a passion which proved unquenchable and which now has burst into flame. This is not the first time I have been fired with love for one of the fair sex, and your husband may well forgive me so long as I do no injury to his honour.' On it goes for quite a bit, until: 'But should I be so happy as to find my love for you reciprocated, then I ask but one favour of you: that I may be permitted to join with you in the procreation of just one little daughter to bear witness of our love – a little Monsonia, through which your fame would live forever in the Kingdom of Flora'. There is no record of Linnaeus receiving a smack in the mouth for his pains, and it is probable that the letter, which was discovered among his papers, was never sent. Nevertheless, little *Monsonia* was duly born and remains a genus even now.

Honoured scientists do not need to hail from biology. Archimedes and Hippocrates can be found in the genera *Archimedes* (bryozoan) and *Hippocratea* (tropical vine), while the species *Lepithrix freudi* (scarab beetle) remembers the father of psychoanalysis. The use of their names is not always arbitrary. Fossilised *Archimedes* bryozoans honour the man's invention by looking exactly like screws made of rock. The bacterium *Janibacter hoylei*, retrieved from the stratosphere between ten and fifty kilometres above earth, was named for the astrophysicist Fred Hoyle, who theorised that life on Earth was seeded by bacteria-like organisms from space.

Many famous people have been honoured with a genus or a specific name, and whether one considers this frivolity or just reward depends on one's view of the person concerned. Few would argue with the spider genus *Plato*, the palm genus *Socratea*, the wasp genera *Lincolna* and *Mozartella** or the goby genus *Zappa*. But should we baulk at the midge *Dicrotendipes thanatogratus* (*thanatos* plus *gratus* equals Grateful Dead), the sand crab *Albunea groeningi* (after Matt Groening, creator of *The Simpsons*), the Mongolian dinosaur *Jenghizkhan* or the land snail *Crikey steveirwini*? In my opinion, if it is acceptable to honour Mozart, then why not the Grateful Dead? Matt Groening deserves all the respect he gets, *Jenghizkan* at least came from Mongolia and Steve Irwin's tribute is neatly appropriate.

The honouring of politicians with Latin names is a tricky business. One man's statesman is another's tyrant, and some namings are designed to condemn rather than celebrate. Inevitably there is *Washingtonia* (a North American Palm genus), *Chesapecten jeffersonius* (fossil scallops) and *Mandelia* (a South African spider genus). But there is also *Anophthalmus hitleri* (a blind cave beetle), *Rotaovula hirohitoi* (a sea snail) and *Kalanchoe salazarii* (a succulent named in honour of the Portuguese dictator). *K. presidentis-malanii* and *K. presidentis-verwoerdii* (named for a brace of South African politicians who devised the apartheid system) were, fortunately, deemed invalid for technical reasons under the botanical code and will trouble us no more. The other

---

* Mozartella does lose some of its gravitas when one considers that it contains just one species: *Mozartella beethoveni*.

side of this coin is shown in Rousseau H. Flower's worm *Khruschevia ridicula* and in the triplet named by Miller and Wheeler in 2005 – *Agathidium bushi*, *A. cheneyi* and *A. rumsfeldi*, all species of slime-mould beetle.

Occasionally, as we have seen, eponyms rise above mere honour (or dishonour) by matching the species or genus named with the person concerned. Terry Erwin, a specialist in carabid beetles, to whom we must be grateful for *Agra katewinsletae*, published many more species of *Agra* in that same paper (he had 'previous' with *Agra vate* and *Agra vation*). The most notorious is *A. schwarzeneggeri*, which he justified as being 'in reference to the markedly developed (biceps-like) middle femora of the males of this species reminiscent of the actor's physique'.[12]

In devising *Psephophorus terrypratchetti*, the author was not merely tipping his hat to the creator of the Discworld novels: *P. terrypratchetti* is a fossil species of turtle, and it is on the back of a giant turtle that Discworld rests. *Campsicnemus charliechaplini* is a species of fly whose back legs assume a bandy-legged appearance on death, and the trilobite *Norasaphus monroeae* was named by Professor Richard Fortey of the Natural History Museum in London for the 'hourglass-shaped glabellum' that forms part of its head. I rather like *Sylvilagus palustris hefneri*, which is, of course, a subspecies of rabbit. But my favourite – everyone's favourite – is *Scaptia beyonceae*, a horsefly blessed with a perfectly round and golden rear end.

Taxonomists clearly need no encouragement in using eponyms (Mark Isaak, to whom I am grateful for supplying many of the above, lists around three hundred on his website).[13] But unless you are a politician, famous pop star, professor or dedicated collector of rare organisms, the chances of being immortalised in Latin are slim. It is, however, possible to buy the right to have something named for you. *Echinofabricia goodhartzorum*, an attractive and woolly-looking sea worm (*Echinofabricia* means 'made of spines'), was named after a 55-year-old maths teacher called Mr Goodhartz. He explained that he did not have children but wished for his name to continue. It cost him five thousand dollars, which seems like a bargain to me.

Large sums of money have started changing hands, and sometimes names, or rather opportunities to name, are auctioned. Most famously, an online gaming company provided the name for a new species of South American monkey, *Callicebus aureipalatii*. *Aureipalatii* means 'of the golden palace', hence the innovative 'common' name for the animal, GoldenPalace.com monkey. It set them back a cool $650,000 and they do not appear to regret their purchase – a new primate species is, after all, a rarity.

The worrying thing for the owners of such eponyms is that names can change and fall into synonymy and so the memorial may not be as long-lasting as the purchaser would like. For science, the selling of names has produced a moral dilemma and the various International Codes have not expressed a view save that they have no view.[14] The funds are welcome in hard-pressed institutes, but stern voices have been raised. Passions run high enough in taxonomic circles without venality raising them further. Scientific integrity could easily be compromised if good species were split into two to 'fulfil an order' for a new name. Taxonomy is a matter of trusting taxonomists as much as it is about evidence, and it is less easy to trust someone who receives money for publication. The Australian herpetologist Raymond Hoser has described the phenomenon of 'cash for names' as 'totally abhorrent'.

There is one unfortunate gentleman who would no doubt have paid good money *not* to have something named after him: Mr Dyar, from whom the moth genus *Dyaria* is derived. Has any taxonomist ever lived to regret bestowing an eponym? Yes, of course. The ichthyologist Rolf Bolin was delighted to receive a signed first edition of one of John Steinbeck's novels from the hands of the author himself. To repay this kindness, in 1939 he named a species of lanternfish *Lampanyctus steinbecki*. Soon after this, and for no known reason, Steinbeck asked for the book back. Bolin spent a great deal of effort trying to synonymise the species with another to rid the world and his slighted pride of *L. steinbecki*, but it remains an accepted species to this day.[15]

Descriptive names

The most sensible names and the most common are those that point out some notable feature of the organism concerned. When used as specific epithets these are very much in the spirit of differentia (see p. 150): i.e. what it is about a species that differentiates it from other members of its genus. For example: *pulverulentus* means 'powdered as with dust', *pulvinatus* 'like a cushion', *pumilus* 'dwarf', *punctatus* 'spotted', *pungens* 'sharp-pointed' and *purpurascens* 'becoming purple'. Every type of description or allusion can be covered in this way: tall or short, edible or poisonous, furry or bald, two-legged or ten-legged, fish-eating or dung-eating, and so on and on.

A black, finger-like fungus that grows on dead beech is called *Xylaria longipes*, from *xulon* ('wood'), *aria* ('pertaining to'), *longus* ('long') and *pes* ('foot'). Hence: 'long-footed thing that grows on wood'. This differentiates it from the more common species *X. polymorpha* ('many-shaped'), which is fatter and lumpier.

Hoplocercus spinosus comes from *hoplon* ('a weapon'), *kerkos* ('tail') and *spina* ('thorny'), translating as 'thing with a spiny tail for a weapon'. Its common name is weapon-tailed lizard.

Mormoops megalophylla, the leaf-chinned bat, comes from *Mormo* ('hideous she-monster'),* *-opsis* ('looks like'), *megalo-* ('big') and *phyllon* ('leaf'). In English that adds up to 'thing that looks like a hideous monster with big leaves'. The leaves are folds of skin around its head that channel sound, and I can confirm that it is no looker. Finally, and in rather poor taste, there is *Scatophagus argus* – the 'many-eyed shit-eater' – whose name comes from *skōr* ('dung'), *fagein* ('to eat') and Argos (after a giant with one hundred eyes).† The latter is a reference to the multiple eye-like spots that cover the body of this Indo-Pacific fish, which is more politely known as the spotted scat.

The above names, on the whole, form useful mini-descriptions of the species concerned. The number of possible combinations, especially when prefixes and suffixes (see p. 77) are added, is endless and can express a huge range of morphological or behavioural features. For example: *Chloephaga* ('shoot eating'), *Bucephala* ('ox-headed'), *Callithrix* ('beautiful hair'), *caligatus* ('wearing boots'), *scabrum* ('scaly'), *auratum* ('golden'), *erythropus* ('red-footed'), *melanura* ('black-tailed') and *griseicapillaris* ('grey-haired').

Sometimes, as in the case of the Australian pratincole, *Stiltia isabella*, a descriptive name is more enigmatic. Admittedly *Stiltia* describes the bird's legs perfectly. But *isabella* is surely an

---

* My much-hated aunt Hilda from Lowestoft comes to mind.
† In fact there is no need to deconstruct this word, because *Skatophagos* is a Greek word for 'shit-eater', although I find the fact that they needed one rather worrying.

eponym? In fact, it is the name of a colour. The creature was described in 1816 by the ornithologist Louis Jean Pierre Vieillot, who was taken by the mottled, dirty brown shade of its plumage. Isabella is a largely archaic English term, frequently found in works on natural history up until the end of the nineteenth century. There are several theories about the word's origin, the best of which involve a dirty woman called Isabella. The most engaging of these is that she was Isabella Clara Eugenia of Spain, who, in a case of domestic solidarity taken too far, swore not to change her underwear until the Siege of Ostend was won. Unfortunately it was one of the longest sieges in history, taking her husband, Albert VII Archduke of Austria, three years to overwhelm the defences.[16] The imagined colour of her underwear at the end of this period matches fairly well that of the Australian pratincole. Sadly, though, if this story is to be believed at all, it must be assigned to the earlier Isabella I of Castile and the less impressive eight-month siege of Granada, because the use of the word predates Isabella of Spain.

Geographical names

Geographical names are a hostage to fortune. As a scientist once complained to me, 'as soon as you name something *mexicanus* or *canadensis*, some bastard finds a specimen of the damn thing ten thousand miles away'. Perhaps this concern explains the extreme hedge-betting of the already vague (it is a fossil fish known only from its teeth) *Asiamericana asiatica*. Despite this drawback, geographical names are as popular as ever, with about one in five new species named after the place where it was found. They are usually noted in scientific papers as 'named for type specimen', which means that no claim is being made that the species is restricted to that location, but that it just happened to be found there first (for more on types, see p. 115). A typical example of a geographical epithet is that of the California horse-chestnut, *Aesculus californica*, a species that has had the decency to remain where it was found.

There may seem to be little opportunity for frivolity in geographical names, but taxonomists always find a way. Excepting when a specific name moves rank to that of genus (see p. 133), there can be no rational reason for naming a genus after a geographical proper name. This, however, has not prevented *Arizona* (a snake genus), *Florida* (bird genus), *Mexico* (beetle) and *Texas* (bug). Occasionally an author will succumb entirely to geographical whimsy and produce such names as that of the braconid wasp *Panama canalis*. One would hope that Linnaeus would have been above such shenanigans, and in this instance he was. His genus *Argentina* is Latin for 'little silver', a good name for the fish that appeared in *Systema Naturae* before the country bearing the same name was even established.

One word of warning about the common geographical epithet *australis*. This and similar constructions mean 'southern' and not, necessarily, 'from Australia'. A plant found only on the south coast of England could quite justifiably have the epithet *australis* (although I know of no example). *Phragmites australis* is the common reed, which perhaps the original author, Antonio José Cavanilles, found only in the south of Spain. In fact, it occurs all over the world and may be among the most widely distributed of all flowering plants.

The epithet *antarctica* may also be misleading. *Dicksonia antarctica* is unequivocally the name for the tree fern, despite it being no secret that tree ferns do not grow in Antarctica. In fact, its nearest location to that cold continent is in Tasmania, sixteen hundred miles to the north. Just as the Greek *arktos*, from which we derive 'arctic', means 'northern', so 'antarctic' can refer to 'southern'. Snakes are even less well suited to Antarctica than tree ferns: the death adder *Acanthophis antarcticus* restricts itself to Australia and New Guinea.*

Habitat names

The seaside is not where most people would go to find fungi, yet on a British Mycological Society foray to Gibraltar Point in Lincolnshire years ago, my companions and I found ourselves transported to anorak heaven on discovering several species nestling in the sand dunes. Among them were the mushroom *Psathyrella ammophila* and the cup fungus *Peziza ammophila*. *Ammophila* derives from the Greek *ammos* for 'sand' and *philos* for 'love' or 'affection', hence 'sand-loving'. The fungi were, of course, surrounded by marram grass, *Ammophila arenaria*. *Arena* being Latin for 'sand', the taxonomist who named it was evidently taking no chances.

Epithets describing habitat are genuinely helpful, unlike many geographic names and most honorary eponyms (which are, like amateur dramatics, engaged in for the delight of the participants alone). They include *silvaticus* ('in woods'), *littoralis* ('on the seashore'), *pratensis* ('coming from a meadow') and *paludicola*

---

* Another commonly misinterpreted specific epithet, although it has nothing to do with geography, is *officinalis*, long used as an epithet for medicinal plants, fungi and even animals and not to be confused with the Latin for 'official', which is *officialis*. It refers to the store room where remedies were kept. *Sepia officinalis*, a name coined by Linnaeus for the squid, is still current today. The ink, called *ossa sepia* by apothecaries, was used, among other things, as an ingredient in ancient toothpastes.

('marsh-dweller'). Habitat is not just a matter of preferred physical geography; it can also be about relationships with other organisms. *Pseudoboletus parasiticus*, for example, is a mushroom that grows on another fungus, the earthball (*Scleroderma citrina*), and *Mussaenda epiphytica* is a cloud forest climbing shrub that grows epiphytically (*epi*: 'upon' and *phyllon*: 'leaf').

Wrong names

Oxymorons, such as *Anoura caudifera* (tailed tailless bat), *Euphoria morosa* (a scarab beetle) and *Unifolium bifolium* (a lily) arise from time to time. Such absurdities are rarely created intentionally or due to lack of concentration; they are the products of taxonomic practice whereby species move genera and bad luck. Examples of straightforward ignorance, however, abound. Linnaeus was particularly productive in this matter, giving us *Lilaeopsis chinensis* (eastern grasswort), which grows in North America, *Lagerstroemia indica* (crape myrtle), which comes from China, and *Scalopus aquaticus* (a mole that is entirely terrestrial).

One very popular misnomer among collectors of taxonomic pottiness is *Chaeropus ecaudatus*, the pig-footed bandicoot. *Chaeropus* means 'pig-footed' and *ecaudatus* means 'without a tail'. The creature was discovered during an 1836 expedition to the Wellington Valley, Australia by Major Thomas Mitchell, who noted: 'This animal was of the size of a young wild rabbit and of nearly the same colour, but had a broad head terminating in a long, very slender snout, like the narrow neck of a wide bottle; and it had no tail'.

This original specimen was evidently dispatched and found its way to the Sydney Museum. Mitchell showed a drawing of it to his friend, a Mr Ogilvy, who, based on this and Mitchell's testimony, published the new species in the *Proceedings of the Zoological Society* as 'belonging to a new genus closely allied to *Perameles* [long-nosed bandicoots] but differing in the form of the forefeet, which have only two middle toes resembling those of a hog, and in the total absence of tail'.[17] The problem is that the pig-footed bandicoot most definitely does have a tail, and a fine one it is, too. The most charitable explanation is that Mitchell's specimen had lost its own tail to a predator. Sadly the creatures were none too good at escaping predators, and, despite the occasional unconfirmed sighting, the species is now considered extinct.*

A similar fate befell *Hydrangea serratifolia*, whose specific epithet translates as 'with serrated leaves'. The species was brought back from the Chilean archipelago by Darwin. Evidently Darwin did not mention to William Hooker, the botanist who named it, that the sample had been nibbled by pests. *H. serratifolia* has entirely smooth-edged leaves.

Taxonomy is challenging at the best of times, but the finest opportunities for making a complete hash of things are afforded by palaeontology. Invitations to err lay temptingly all around like bright jewels. I have a great deal of sympathy and admiration for palaeontologists. They must work with scarce specimens, random samples of organisms from the past, which are often in very poor

---

* The mistake can, perhaps, be forgiven because many four-legged animals do indeed lack a tail. No bird, however, has ever been discovered that has no legs. Early European naturalists, examining the skin of the bird of paradise, noticed that it was *sans* legs, which of course had been cut off during the skinning process. Evidently this possibility had never occurred to them; Linnaeus followed current understanding and named the creature *Paradisaea apoda*, 'legless bird of paradise'.

condition and usually missing large chunks. Indeed, the naming process itself reflects the difficulties involved, because it is common for a single organism to possess several names. A researcher may describe the tip of a therapsid's nose one year and the tip of the tail from the same species the next year. The nose and the tail will have different names. *Stigmaria* and *Lepidostrobus*, for example, are respectively the fossil roots and fossil cones of the tree *Lepidodendron*.

Of all the disasters and embarrassments that litter the history of palaeontological taxonomy, none is more egregious than the story of *Aachenosaurus multidens*. Moresnet near Aachen is a town balanced precariously on the borders of Belgium, Holland and Germany. It is now a German town, but in 1888 it was neutral territory. That was the year in which Gerard Smets, a doctor of natural sciences, professor at the College Saint-Joseph in Hasselt and keen amateur palaeontologist, discovered in a quarry at Moresnet two small fossils that would change his life. He examined his finds, decided that they were fragments of the jawbone of a dinosaur previously unknown to science, and published his findings, naming his dinosaur *Aachenosaurus multidens* – the 'Aachen lizard with lots of teeth'. In so doing, he established not only a new species, but also the type species of a new genus. Heady stuff for a non-specialist.

Now, I have been studying the larger fungi for fifty years. While I am not up there with the professionals, I do know a fair bit about them. But if I found a mushroom that I thought was a new species, and ten times more so if I thought it a new genus, I would seek the opinion of every mycologist I knew before offering a paper for publication. Perhaps Smets had no friends on whom to call, or perhaps he was blinded by the prospect of glory, but either way his paper was printed, unread by anyone who could have saved him from the humiliation that lay in wait.

Smets may have lacked friends or modesty, but imagination at least he possessed in abundance. There was only a small amount of material from which to draw conclusions, but such paucity of evidence was capable of deterring only lesser men and he managed to paint a remarkably detailed portrait of his dinosaur. He assured himself of the bony nature of the fossils by studying them with a lens and a microscope, and also observed some incrustations, which he identified as teeth. He concluded that the dinosaur was of the order Ornithopoda ('bird-footed') and family Hadrosaur ('bulky lizard'), a group of dinosaurs with broad, flat jaws that they used for scooping up water plants. Going further, Smets stated that it walked on two legs, would have been four to five metres long, was covered in spines and had a spoon-shaped jaw with which it ate its favourite food – soft vegetation. And, of course, it had lots of teeth.

M. Smets's nemesis came in the much respected form of one Professor Louis Dollo, a professional palaeontologist from the Royal Belgian Institute of Natural Sciences in Brussels. Professor Dollo carefully examined the fossils of the putative *Aachenosaurus* and came to the inescapable conclusion that they were pieces of petrified wood.

Any feelings of sympathy for the unfortunate, if hopelessly inept, Smets would be ill-placed. Instead of hanging his head in shame, thanking Professor Dollo for his guidance and vowing to be a damn sight more careful next time, he came out, guns blazing, with what was described in a letter to the *Geological Magazine* as a 'crushing retort' designed more to accuse than refute. He viciously attacked Dollo in print and attempted to cast doubt on the Professor's abilities by citing palaeontologists who had disagreed with him in another context.

His accusation was rebuffed by those same scientists, G. A. Boulenger and Richard Lydekker, in an article pointedly titled 'The Wooden Dinosaur', which noted that Smets's response 'followed the Old Bailey maxim, that when you have no case, the only thing left is to abuse the plaintiff and all connected with him'. When Dr Maurice Hovelacque from the Geological Society of Paris agreed with Dollo's assessment in his article 'On the vegetable nature of *A. multidens*', it marked the end of the dinosaur that had never lived. A letter to a highly embarrassed journal associated with Smets and his dinosaur began an article with: '*Horrible! Horrible!*' (you will need to imagine this in a French accent to get the full benefit) and went on to suggest that the society distance itself from Smets.

Finally accepting defeat, Smets, after publishing a single further paper in 1889 on the much safer subject of tortoises, gave up science altogether. The humiliation of his encounter with the world of palaeontology was too much for him and he retired to obscurity. However, *Aachenosaurus multidens* (a *nomen dubium*, see p. 136) lives on at the top of alphabetical lists of dinosaurs, a prominent and perpetual memorial to scientific hubris.*

To be fair to Smets, his *nomen dubium* was a mistake. But at least one has been created mischievously. In 1975, in an article for

---

* The honour of being at the top of the alphabetical list of organisms goes jointly to *Aa*, a mollusc, and *Aa*, an orchid. Fulfilling the rule that names must contain at least two letters, they are in unassailable positions. The imaginary genus *Aaa* looks more impressive but must come later. Unlike first place, there is no theoretical last place beyond an infinite string of 'z's, but the current record holder appears to be *Zyzzyzus*, a genus of tiny marine invertebrates related to jellyfish.

Nature, Sir Peter Scott coined the binomial *Nessiteras rhombo-pteryx* for no less a creature than the Loch Ness monster. It translates as 'the Ness monster with diamond-shaped fin'. At that time there did seem to be some chance that the beast existed, and the naming of it purportedly afforded the creature conservation status – at least it did until the Conservative MP for Kinross and Perthshire, Nicholas Fairbairn, pointed out that the binomial was an anagram of 'Monster hoax by Sir Peter S'.

Plain silly names

Anagrams have been popular since Linnaeus's day. The great man himself accepted the genus *Alcedo* for kingfishers from the Latin, but as more kingfishers were described, the genus needed to split and the new genera needed new names. What better than to call them *Lacedo* and *Dacelo* (kookaburras)? Similarly, *Eleotris* and *Erotelis* are both genera of sleeper goby (fish), while *Ptinus, Niptus* and *Tipnus* are all beetle genera. Inevitably, less innocent anagrams have been produced. The Figueiredo paper mentioned earlier related the story of the tropical plant *Kalanchoe mitejea*. It was named by the eccentric twentieth-century botanist Raymond-Hamet in collaboration with his friend Alice Leblanc; *mitejea* is an anagram of *je t'aime*, clearly indicating the nature of their friendship. Raymond-Hamet also coined *Sedum leblancae* and *S. celiae*, the latter being a partial anagram of Alice. I think this is rather romantic,* although if there was a Madame Raymond-Hamet, she presumably would not have agreed.

---

* He had form in this, because he was well known for naming species after ladies of his acquaintance, the succulents *Kalanchoe adelae* for Madame Adele Le Chartier and *K. luciae* for Mademoiselle Lucy Dufour being further examples of his largesse.

Sometimes, when taxonomists are unable to point to anything about an organism that distinguishes it, or unable to convince even themselves that something is really a new species, they throw up their hands in dismay and descend into prevarication. Thus we have a genus in the Asteraceae (daisy family) called *Perplexia*. There is also a sweet box called *Sarcococca confusa*, presumably to distinguish it from half a dozen others in the genus that were not confusing. The most common of these weasel epithets is *dubius* and its variations. On a quick search, I unearthed seventy-seven such names, revealing a wholesale dereliction of care.

Unhelpful as these names may be, each one does at least tell us something about the species concerned – that it is difficult to distinguish from its sister species. However, no information at all is conveyed by the names listed in this imaginary (and nonsense) news headline: 'Extra! Tuxedo Saga Bishops Box Circus Chorus Disaster!' Every word is a genus name. There are perhaps one hundred such seemingly random words used as generic names.

The species *Abra cadabra* (now regrettably synonymised as *Theora mesapotamica*) was clearly irresistible when Eames and Wilkins attached it to a fossil bivalve in 1957,[18] but at least it was based on an established genus and a pun waiting to happen. However, sometimes genera are named with the express intention of engaging in punning. In 2002, Neal Evenhuis devised *Pieza*, a genus of fly, and proceeded to describe *P. kake*, *P. derisistans*, *P. pi* and *P. rhea*.[19] H. M. Woodcock, in 1917, gave us the heterotrophic flagellate *Kamera lens*, perhaps because it sports a single round eye or eye-like structure, or perhaps because he thought it sounded cool.[20] No doubt the challenge of dealing with large numbers of difficult taxa, which lack distinctive features, is the reason for such imaginative names, which in turn explains why entomologists are the worst offenders. Once this sort of thing begins, the floodgates open to allow the likes of *Eubetia bigaulae* (a moth), *Heerz tooya* (a wasp), *Ba humbugi* (a snail), *Eurygenius* (a genus of flower beetles), *Gelae baen, Gelae belae, Gelae donut, Gelae fish, Gelae rol* (all fungus beetles) and *Apopyllus now* (a spider), although I have to admit that last one is pretty good.

Just to show that silly names have a long pedigree, behold Linnaeus's name for the hoopoe, *Upupa epops*, an onomatopoeic name derived from Aristophanes's play *The Birds*, Epops being the name of the hoopoe in the play. I have always loved the name for brooklime, *Veronica beccabunga*. To my ears, it is reminiscent of Carmen Miranda, and I imagine them singing duets.

Mark Isaak lists many rhyming, palindromic and oxymoronic names. Some are accidental, but others are due to taxonomists having fun: *Apus apus* (the swift), *Cedusa medusa* (a bug), *Semicytherura miii* (the tiny crustacean with more than its fair share of 'i's), *Aiouea* (a vowel-rich genus of laurels) and the rather silly *Brachinus aabaaba* (a ground beetle).

Generally these frivolous names are accepted with a wry smile by the scientific community. But some scientists do not like them at all. In a 2007 paper, the aforementioned herpetologist Raymond Hoser, reflecting on the mite genus *Darthvaderum* and others like it, noted: 'assuming the species named is a valid and previously unnamed taxon, we have to accept the names and use them, even though we may cringe every time we do so'.

The cringing is about to get worse.

Rude names

Years ago, I ran a series of Saturday fungus forays and lectures for my local adult education authority. I would lay out our finds on a plain Formica table and write their Latin names underneath using a whiteboard marker. One week, I used a permanent marker by mistake, so I folded up the table and put it at the back of the stack in the hope that no one would find it before the following week, when I would bring in a cleaning solvent. On the Wednesday, my principal called to tell me of the 'terrible complaint' she had received from an agitated elderly gentleman from the art class, who had been given a table covered in 'filthy words'. I checked my list for that day. Among the thirty or so species we had discovered were *Phallus impudicus* (common stinkhorn), *Amanita vaginata* (grisette), *Clitocybe nebularis* (clouded funnel) and *Geastrum fornicatum* (arched earthstar). You could see the man's point. This is a barely credible story, but you have my word that it is true.

I apologise. I am going to say this only once, because if I do so every time I use a rude word or allusion I will never get to the end of this chapter. I will be as polite as is possible, but I can only do so much. If this subject is likely to be too distressing for you, then turn immediately to p. 62. If you think that I am bringing taxonomy into disrepute, I remind you that I did not create any of these names. Each one was devised by a taxonomist.

I could have called this section 'Carry on Taxonomist'. It has no deep significance but is simply adolescent and sometimes kindergarten humour: unworthy, superficial . . . and funny. Scientific names that you would not repeat to your mother are so common that one must suspect taxonomists of creating them intentionally. Sometimes they do, but more often it is a matter of innocent double entendre and poor pronunciation, wilful or inadvertent.

As discussed, many Latin names helpfully describe morphology, using terms that may be familiar from anatomy. One often comes across *vaginae*, for example, and if an organism looks like a *phallus* the taxonomist is only doing his duty in saying so.* Nomenclatural vulgarisms have a long pedigree dating back to Linnaeus and beyond. But it was Linnaeus who, as with so many things, set the standard. In the coining of vulgar binomials, he was the master.

First, the unfortunate matter of the suffix *-anus*. In Latin nomenclature, it simply indicates position, connection or possession by, as in *sylvanus* ('belonging to woods'), *africanus* ('coming from Africa') and *alphonsianus* (for Professor Alphonse Milne-Edwards). It has nothing to do with anything anatomical. The English name for the body part is from the Latin noun with the same meaning, itself a derivative of *anulus* or *annulus* – 'a ring'.

On the page, or if pronounced as in 'pat' or in 'part', the suffix is unremarkable; only when pronounced (properly, as it happens) with an 'ay', as in 'pane', does it become a source of infantile sniggering (for notes on pronunciation, see p. 62). Taxonomists, wary of offending those they wish to honour, tend to avoid the suffix, if possible. But some names suffer more than others, and no doubt Milne-Edwards was delighted with his epithet. Professor Roy Watling, formerly of the Royal Botanic Garden Edinburgh, told me of the occasion he and his colleague Alex Smith wished to name a new species of mushroom in the genus *Leccinum* (a bolete).

---

* Although quite what William Elford Leach was thinking when he changed Linnaeus's perfectly respectable name for earwig from *Forficula minor* to *Labia minor* I have no idea.

It was to honour the distinguished boletologist Walter Snell. Faced with the unthinkable *snellianus*, they settled on *Leccinum snelli*.

Other taxonomists have not been so considerate. The nineteenth-century botanist William Hemsley, for example, does not appear to have thought through his name for the bramble species *Rubus cockburnianus*, with which he wished to honour the Cockburn family. Rafinesque, although of French descent, lived and worked in the US, so he should have realised that his *Soranus* was open to misinterpretation. More forgivable because of the language difference is *Bugeranus*, the generic name of *B. carunculatus*, the wattled crane, with which the German ornithologist Gloger presented Herr Buger. For temporal as well as language reasons, the nineteenth-century German botanist Karl Sigismund Kunth cannot be held responsible for the specific epithet of the invasive gamba grass *Andropogon gayanus*, with which he honoured the French botanist Claude Gay. However, P. J. Hancox, writing in 1987, must be guilty as charged for giving us the improbable imperative therapsid genus *Dolichuranus*; that *dolichos* is Greek for 'long' does not forgive.

I have canvassed every one of my mycologist, botanist, phycologist and zoologist friends and acquaintants for amusing vulgarisms. It appears that the one most often repeated in pubs after a field trip or the publication of a new paper is the indecent proposal *canaliculata*. It means 'channelled', a common morphological feature, and as such enjoys currency in most disciplines. There is *Pelvetia canaliculata*, the channelled wrack; *Pomacea canaliculata*, the channelled applesnail; *Puccinia canaliculata*, a rust fungus of cypress trees and, best by far, the channelled heath, *Erica canaliculata*. Another collectors' item is *Hornia*, a genus of blister beetles parasitic on bees, which is named, I presume, for a Mr Horn. On its own it is a rather pedestrian pun, but is raised to worthiness by one of its species: *Hornia minutipennis*. *Pennis* refers to wings, of course, but let us not quibble.

Amanita vaginata is an attractive, edible fungus that I find frequently on my fungus forays. Many *Amanita* species have a universal veil from which the fruiting body (the mushroom itself) emerges. In *A. muscaria*, the fly agaric, for example, it adheres to the cap and breaks up to produce the familiar white spots. In *A. vaginata*, however, it retains its integrity and remains as a sheath at the base of the stem. The Latin for 'sheath' is *vagina*, a word now used to describe the female birth canal, as if you didn't know. I am not one for sexual politics, but 'sheath' is defined as a close covering, often of a dagger or sword, and I wonder what this organ would have been called had it been named by a woman. Probably *canalis*, indicating its productive rather than recreational function. Being a common morphological structure, *vaginae* are found all over the place, for example in the grass *Festuca vaginalis*, the branching vase sponge, *Callyspongia vaginalis*, and the algae genus *Vaginaria*. The latter is particularly troublesome because it includes *V. vaginata* and *V. vulgaris* and the original author of *Vaginaria* is Kuntze.

Otto Kuntze was a great nineteenth-century German botanist and mycologist who blessed many a species with his easily mispronounced author citation (I understand that 'Koontze' gives the proper feel of it). Alone he was fairly harmless. Unfortunately he had a contemporary in the person of the mycologist Karl Wilhelm Gottlieb Leopold Fuckel. Kuntze revised one of Fuckel's species, and mycology is now further blessed with *Cucurbitaria applanata* (Fuckel) Kuntze. Many author names are abbreviated (see p. 113), but English-speaking taxonomists have resisted the temptation to abbreviate Fuckel. Not so non-English-speaking taxonomists, who have occasionally but famously provided us with 'Fuck.' in their citations.[*]

The subject of vaginae brings me on to a passage in Linnaeus's travel journal of his 1741 journey to Gotland and Öland, which I have edited slightly for clarity if not decency. 'We marvel at the shell *Cunnus marinus* . . . in which Nature has portrayed the female genitals; Nature here has no less art in portraying it in all its detail'.[21] The blame for *Cunnus marinus* cannot be laid at Linnaeus's feet because it appears in Bauhin, who himself ascribed it to earlier writers. It is absent from modern lists, being considered a pre-Linnaean name.

The Latin word *vulva* is for female sexual organs in general, although in English it refers to the external genitalia. Occasionally the word appears in scientific names in its other form of *volva*, as in *Volvariella*, a genus of mushroom very similar to *Amanita* in having a sheath or *volva* at the base of its stem. However the main use of this word in nomenclature is for something quite different. Botanists and mycologists have long faced the problem of describing certain odours in ways that are immediately recognisable to their readers. Two common smells have proved particularly challenging, but both have been met with commendable directness, and one of them moderated by some gentle Latinising. The first is a smell found frequently in fungi, that of semen. It is usually referred to as 'spermatic',[†] but does not appear to have

---

[*] A more innocent example of humorous author citation is the butterfly genus *Petula* Clark.

[†] 'Mealy' is often used as a mealy-mouthed euphemism.

been used in naming. The second is the odour of human female genitalia, which comes partly from the chemical trimethylamine.[22] This is produced by bacterial action on vaginal secretions and is, I am told, a natural attractant for the male. Among the many goosefoots is the stinking goosefoot, Linnaeus's *Chenopodium vulvaria*. A distinguished English botanist, who will remain nameless, is reported to have said: 'It doesn't say much for Frau Linnaeus.'

There has been little evidence of subtlety so far, but oblique nomenclatural references to unmentionable parts can be rather endearing. Linnaeus was a master of this art, as can be seen in his *Venus dione*, a remarkably beautiful bivalve from the West Indies. In *Fundamenta Testaceologiae* (1771) Linnaeus included a drawing of the creature, displaying its glory. From the helpful angle Linnaeus presented, the animal bears an uncanny resemblance to external human female genitalia (you really need to find a picture). In case anyone missed the point, he labelled various parts as vulva, labia etc. The belt-and-braces name he gave it, *Venus dione*, reflects this unequivocal imagery: Venus is the goddess of love and sex, and the purportedly beautiful Dione is the mother of Aphrodite, the Greek equivalent of Venus. This species no longer resides in *Venus*, but the sexual reference lives on in its new genus, *Hysteroconcha* (Greek *hystero* for 'womb' and *konche* for 'shell').

The same imagery is found in one of the most extraordinary members of the plant kingdom: the Venus flytrap, *Dionaea muscipula*. Here, both goddesses are referenced, one in the common name and one in the Latin. The reason for this was made clear to me during a pleasant pub meal in the company of a group of insectivorous plant enthusiasts. (They displayed all the single-minded eccentricities one might expect, although they were no odder than the mycologists I am used to.) One of them, Tim Bailey, told me the story of the Venus flytrap and was kind enough to give me a copy of his charming book on this plant, which told all.[23]

In 1759, Arthur Dobbs, the Governor of North Carolina, where the Venus flytrap is found, wrote to the naturalist Peter Collinson, describing the plant and its extraordinary method of catching flies. The Venus flytrap quickly became a 'must-have' species. An indication that the plant was interesting for more than its fly-catching abilities is found in a letter from another naturalist, John Bartram from Pennsylvania, who refers to it as a 'waggish plant, waggishly described', although the 'waggish' description itself appears to be lost. In some of the letters the Venus flytrap is called Tipitiwitchit, but with no accompanying explanation. However the true meaning is revealed in a letter from Bartram to Collinson, which discusses a third naturalist (Dobbs), on whom the supply of plants was dependent: 'I hear my Friend Dobbs at 73 has . . . married a young lady of 22 [note: Justina Davis, who was actually 15]. It is now in vain to write to Him for seeds or plants of Tipitiwitchit now he has got one of his own to play with'.

Muscipula means 'fly trap', and it has been unconvincingly suggested that *dionaea* is a reference to the beauty of the plant. But beauty it lacks, and there is little doubt that it is the red, moist, gaping, hair-fringed leaves that provided the inspiration.

The genus *Clitoria* appears in Linnaeus's *Species Plantarum* (1753), although Linnaeus cannot be held entirely responsible for this one either, because Jacob Breyne in 1678 had the species *Flos clitoridis ternatensibus*.[24] Linnaeus provided few etymologies, so it has been left to commentators to tell us that this member of the pea family is 'probably' named for the shape of its flowers. There is no 'probably' about it.

Breasts make a frequent appearance in nomenclature, most frequently in mycology. This is not due to any suspect preoccupation among mycologists. It is simply because most mushrooms are shaped at least a little bit like breasts. The parasol species *Macrolepiota mastoidea* looks distinctly like a breast, because it has a little bump in the centre of the cap that is reminiscent of a nipple. The epithet is from the Greek *mastos* for 'breast'.

Very nearly my favourite rude name is *Inocybe eutheles*. Almost beyond question it was a joke between its author, the famous English mycologist M. J. Berkeley, and his associate C. E. Broome. The species is creamy buff in colour, distinctly rounded and with a little point on the tip of the cap. *Eu* means 'true', or 'nice', and *theles* is Greek for 'teats'. Few would argue that *eutheles* means anything other than 'nice tits'.

Among the fungi on my classroom 'show and tell' table was the remarkable stinkhorn, *Phallus impudicus*. If you have ever seen one you will know that its name is entirely deserved: *Phallus* derives from the Greek *phallos* for 'penis' and *impudicus* is Latin for 'lewd' or 'unashamed'. A similar and related fungus within the order Phallales is *Mutinus caninus*. *Mutinus* is the Roman name for the Greek god of reproduction, Priapus, and is often used to mean 'penis'. So the fungus's literal translation is 'dog's penis', although the common name dog stinkhorn puts it more politely.

Phalli are as popular in nomenclature as *vaginae*. For example, *Phallosoma priapuloide*, a marine worm, manages to squeeze two penises into one binomial (*Phallosoma* means 'penis body'). Two sea squirt genera are inevitably blessed with penis references: *Phallusiopsis* ('looks like a penis') and *Phallusia*. I often find a member (sic) of the latter genus when I go shrimping and clam hunting off the Dorset coast. They are white, a standard British six inches long and covered with little round bumps. If you squeeze one it squirts seawater out of its end. One of the fellows I go shrimping with is a Portland crab fisherman who calls them, straightforwardly enough, piss cocks. Not to be outdone, I told him that the Latin name, *Phallusia mammillata*, means 'penis covered in breasts', although, to be honest, those weren't quite the words I used.

Occasionally on my coastal forays I find the slipper limpet *Crepidula fornicata*. This species often forms 'breeding stacks', one on top of the other, and one would imagine that this wanton behaviour has provided them with their name. The earthstar fungus *Geastrum fornicatum*, however, shows no immoral tendencies, so its name appears to be a mystery. In fact, *fornicata* and *fornicatum* are entirely innocent terms for 'arch' or 'arched', from the Latin *fornix*. The shell of the slipper limpet is curved like an arch, while the 'rays' of the earthstar arch backwards to lift the puffball-like central body of the fungus above the woodland floor. Indeed, 'fornication' itself also derives from *fornix*, as Henry Mayhew explained in *London Labour and the London Poor* (1851): 'The *fornices* of Rome were long galleries, divided into a double row of cells – some broad and airy, others only small dark arches, situated on a level with the street and forming the substructure of the houses above . . . In these long lines of cells the prostitutes of the poorer class were accustomed to assemble, and thence was derived the ecclesiastical term fornication, with its ordinary English meaning'.

Similarly, the specific epithet *urinatrix* brings all sorts of unsavoury images to mind, but, like *fornicatum*, it is quite innocent. It appears in the name of the common diving petrel, *Pelecanoides urinatrix*, whose generic name simply states that it 'looks like a pelican'. The offending specific epithet derives from the Latin for 'diver' which is *urinator*. 'Urine' comes from the same root, *ūrīnāre*, 'to plunge underwater'. *Urinator* was once given to a genus of loons (divers), but fortunately this burden has been lifted and they are now known by the more respectable name of *Gavia*. Less fortunately, we appear to be stuck with the wasp genus *Pison*, established in 1808 by the Swiss naturalist Louis Jurine. Jurine is blameless in this story because the name may refer to Pishon, one of the rivers mentioned in Genesis, or possibly the Greek word for 'pea', although neither theory is particularly convincing. With such a tempting name waiting for an unsuitable specific epithet, it was only a matter of time before someone fell from grace. That someone was the American entomologist Arnold Menke, who in 1988 named a new species *Pison eu*.

To counteract the impression that all taxonomists are corrupted by impure thoughts, it is worth recording the relatively chaste genera of bugs created by G. W. Kirkaldy in 1912. Among them are *Pollychisme*, *Marichisme*, *Dolichisme* and *Florichisme*. Kirkaldy was condemned by the Zoological Society of London for these names, which, considering the antics of his fellow scientists, seems a little unfair.[25]

A note on pronunciation

Some of the examples given above would not seem nearly so funny if we knew how to say them properly.

The significance of pronunciation was made clear to me a quarter of a century ago, during a trip with the British Mycological Society to the Italian town of Alba. This being the home of the white truffle, *Tuber magnatum*, it is the only place to be in October if you are a mycologist or a gourmet, and I fancied myself as both. Among our group was Italy's leading *Tuber* authority, Dr Giovanni Pacioni from Perugia. I asked him, during an expedition to the hills around the town, if we were likely also to come across the summer truffle, *Tuber aestivum*. Dr Pacioni's English was very good, but he looked at me, puzzled, and asked if I would repeat the question. I did, several times. 'Yes, *Tuber aestivum* – will we find any?' and '*Tuber aestivum*, are there any around here?' Then '*Tuber aestivum, Tuber aestivum*?', followed by a certain amount of hand waving.

Dr Pacioni was a specialist in the genus *Tuber*, so it seemed unlikely that he had never heard of the commonest *Tuber* species. Eventually, however, his eyes brightened and he said 'Ah! You mean *Tuber aestivum*!' But of course he said it differently. My version was '**east**-i-vum' (the stress falling on the letters rendered in bold type). His was 'est-**eave**-um'. Well, he was Italian and therefore a direct heir to the language of Rome, so there seemed to be little doubt that he was right and I was wrong. Just to rub it in, he told me that the British were the very worst at pronouncing Latin binomials. So now you know.

But was I really wrong? Pronunciation is more a matter of usage and convention than of rules or being faithful to how people spoke a particular language two thousand years ago. Scientific Latin is a language all of its own, with its own pronunciation. Among English speakers at least, it follows, more or less, what is known as 'traditional English' Latin pronunciation. This contrasts quite starkly with 'reformed academic' pronunciation, which seeks to approximate the pronunciation of the average educated ancient Roman, and also with 'Church Latin', which is different again. According to British convention, my pronunciation of *Tuber aestivum* was right, but it can hardly be said that a highly educated Italian such as Dr Pacioni was wrong. He was just right as well.

I meet many people who are new to mycology and botany, or who have read hundreds of binomials without ever being required to say them aloud, so I am often asked how to pronounce things. Very often people get excited to hear me pronounce something they have always wondered about. The mushroom genus *Clitocybe* ('cly-toss-**eye**-bee') is a frequent hit – 'Oh, *that's* how you say it; I have always said "cly-toe-sybe" '. There are rules of pronunciation both in terms of how individual letters should be rendered and on which syllable the stress should be placed. But there is enormous variation in how people actually say things and even the many published guides to scientific pronunciation sometimes fail to agree.

People are actually better at Latin pronunciation than they think, owing to the common use of genus names in horticulture and classically derived words in medicine, science and elsewhere. When considering an unfamiliar name, it is worth recalling vaguely similar words, such as chrysanthemum, peony, agapanthus, ceanothus, appendectomy, oesophageal and seismology. These few words alone encompass a large number of 'rules'.

Learning to pronounce any language by following a set of rules is not unlike learning to drive by taking a car to pieces. Also, and unfortunately, rules alone will not always provide the correct pronunciation, as it is often less than obvious when a letter should be long (as in 'me' and 'mate') or short (as in 'met' and 'mat'), a matter which also affects if and where the stress should be placed. For this the stem word must be found in a classical dictionary and the length of the letter determined; a process fraught with difficulty.

Written Latin is a mercifully phonetic language. There are no silent letters or nasty surprises, such as the -gue at the end of the borrowed French word fatigue, which caused me so much embarrassment in class when I was ten. It also explains to anyone who doesn't yet know (as I once didn't) what *Cotoneaster* has to do with Easter, which is nothing at all: it is pronounced 'ee-aster'. Every letter is enunciated, with three exceptions. The first is diphthongs, of which there are six: *ae*, pronounced 'ee'; *au*, pronounced 'aw'; *ei*, pronounced as in 'height'; *eu*, pronounced 'yew'; *oe*, pronounced 'ee'; and *ui*, pronounced as in 'ruin'.

A second exception is the proper names included in binomials that celebrate people or places. The specific epithet *cartwrightii* commemorates someone called Cartwright, and just because it appears in Latinised form, there is no reason why a (doomed) attempt should be made to pronounce it phonetically. While someone may be able to pronounce *cartwrightii*, they may be completely stumped by *Warszewiczella* (or, of course, the other way round), so people must muddle through as best they can. The third exception is the occasional Greek word with a silent letter lurking within. The most obvious is *Pterodactylus*.

Vowels follow English and can be long, as in 'pike', or short, as in 'pick'.

Most consonants are used in the familiar manner, too, so only a few are worth mentioning:

— *c* is hard (as in 'cat') in front of *a*, *o*, and *u*, but soft (as in 'century') before *ae*, *e*, *i*, *oe* and *y*, so *coccus* is pronounced 'kok-kus', but *coccinia* as in 'accident';

— *g* is hard (as in 'garter') before *a*, *o* and *u*, but soft (as in 'germ') before *ae*, *e*, *oe*, *i* and *y*;

— *ch* is, with rare exceptions, always hard, as in 'chord';

— *s* is used as in 'section', unless it comes between two vowels or after a consonant at the end of a word, in which case it sounds more like 'z' (as in 'rose').[26]

Beyond this it is my intention to remain almost silent, because the task of teaching pronunciation with rules is quite impossible and any attempt would be too long and extremely boring. It would also be pointless since, as noted above, the rules are more what you would call 'guidelines' than actual rules. However, I would like to describe how to pronounce a notoriously problematic word ending: *-oides* (see p. 83), as in *Hericium coralloides*, a fungus that 'looks a bit like' a coral. The *oi* is frequently pronounced 'oy', as in 'soya', to give 'oy-deez'. But *oi* is not a dipthong, so the vowels must be enunciated separately – 'oh-i-deez' with a short 'i'. Other endings that cause problems include *-ii*, as in *smithii*, usually rendered 'ee-eye' (you are on your own with the specific epithet *miii*), and *-aceae*, the family name ending, as in Agaricaceae, which is 'ay-see-ee'.

Should you ever face the necessity of pronouncing Latin names in public my heartfelt advice is to use straightforward intimidation and do so VERY LOUDLY AND WITH ENORMOUS CONFIDENCE. If they know of the organism concerned they will more than likely know what you mean and will assume it was they who had always mispronounced it in the past.

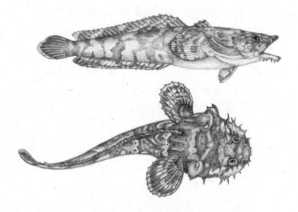

Batrachoides pacifici

CHAPTER III

The LANGUAGE of NAMING

UNLESS MY LATIN TUTOR, DR PARKER, HAS REACHED THE AGE he always looked (one hundred and twenty), he will never know that I have taken it upon myself to explain the details of Latin grammar to innocent third parties. Nevertheless I feel his stern expression turned towards me as I write, and I quake.

A scientific name consists of two Latinised words in the general form of noun and qualifier – as in 'brown duck', 'Johnson's lizard' or 'elephant shrew'. The genus (duck) is the noun and the specific epithet (brown) the qualifier.

The generic name is usually a noun (*Sorex*: 'shrew') or several words that may or may not be nouns joined together to look like one (*Glycyphagus*: 'sugar eater'), or, at the very least, a word in which the noun is implicit, as in adjectivally inclined names, such as *Scabiosus*: 'rough [thing]'. Completely invented names, such as anagrams, automatically achieve the status of nouns. For example, the crustacean genera *Conilera*, *Lironeca*, *Nerocila* and *Olencira* are anagrams of Caroline, the wife of their author, W. E. Leach. Since these four words were first coined as names for genera, they must, obviously, be nouns.

The specific epithet may take the form of an adjective and may indeed be an adjective (*vulgaris*: 'common'). However, it can

be an adjective–noun combination (*megalophylla*: 'big leaf').
Adjective–adjective and adjective–adverb combinations are also
used, although it is often hard to tell them apart. *Acutesquamosus*
means 'acutely scaly', but easily could be 'acute scaly'. There can
also be a participle pretending it is an adjective (*imitans*:
'imitating') and several other adjectival arrangements.

Possessive specific epithets can be found in such names as
Sorex granti ('shrew of Grant'), *Colocasia antiquorum* ('water
plant of the ancients') and *Acmena smithii* ('acmena of Smith')
– although for once I much prefer the common name, which is
lilly pilly.

A specific epithet can also be a noun, as in our very own *Sorex
araneus*: 'spider shrew'. A noun used in this way is called a noun in
apposition.

As with genus names, there are oddities. Anagrams, nonsense
words, verbs and even imperative clauses – such as in *Impatiens
noli-tangere*, where *noli-tangere* means 'do not touch' – are
sometimes used.

Inflection

As already noted, the two words that make up a binomial are
required to be grammatically correct. You cannot just take two or
more words from a Latin dictionary, put them side by side and
expect them to make sense – they need to be appropriately
inflected. This, of course, is also true of English, which is slightly
inflected, but not remotely to the same degree. English was cast
from the melting pot of Anglo-Saxon and Norman French, and
most of the inflections were mercifully boiled away. Latin is a
highly inflected language, meaning that the relationships between
words are determined by word endings rather than word position.

If the words that make up a Latin sentence were placed in a bag and shaken, you would stand a very good chance of understanding the sentence when they were emptied onto the table – assuming, of course, that you understood Latin in the first place. Try this with English and you will get nowhere.

A word ending depends on case, gender, number and declension. There are six standard cases in Latin, although mercifully only the nominative and genitive are used in biological nomenclature. A word in the nominative is a noun that is the subject of a sentence or phrase, and a word in the genitive case marks possession. 'John's cat' is a simple example, where 'cat' is the nominative and 'John's' the genitive. Like many languages, Latin is afflicted by gender and gender endings for objects and concepts that display no sexual identity whatsoever (this disease is largely absent from English but in Latin achieves its most virulent form). Nouns can be masculine, feminine or neuter. Number is, of course, an inflection familiar to English speakers. But by far the worst thing, and the point at which Dr Parker and I parted company, is that word endings vary according to which of five nounal and three adjectival declensions the word belongs. There are further complications with word endings, such as irregular forms and the special attention required by endings for words of Greek derivation.

To demonstrate this in action, here is a simple example of the simplest form of a binomial: noun and adjective. The grey-headed woodpecker is *Picus canus*: 'grey woodpecker'. If we wished to apply this adjective to a feminine noun such as *Rana* (frog), then the ending would have to change to *cana*, giving us the splendid, if imaginary, *Rana cana*.

If the specific name should agree in gender with the generic name, what of the killer whale, *Orca gladiator*? The endings are respectively feminine and masculine and so are not in agreement. This is because they are nouns in apposition and the rule does not apply. Of course, you can have two nouns that are not in apposition, because nouns can be used in the genitive (the parrot of Johnson, where 'of Johnson' is the genitive noun, for example). In this case the specific epithet would have to agree. This fine grammatical distinction has caused endless problems.

A new species of crab was discovered in Sri Lanka in 1992 by Peter Ng. It was in an established genus, *Ceylonthelphusa*, so he just needed the specific epithet. Its extraordinary habit of climbing trees provided him with *scansor*, which means 'climber'. He made it clear in his etymology section that this was a noun in apposition and so he did not have to match his masculine specific name with the feminine genus name. Not everyone is as careful as Peter Ng and not so considerate as to state specifically the grammatical form he is using.

Once a name is published the world is pretty well stuck with it, but grammatical errors can be corrected. A list of bacteria was found to have fallen foul of this pitfall in nomenclature, having treated nouns in the genitive as being nouns in apposition, when clearly from the context they were not. *Pantoea ananas* was duly amended to *Pantoea ananatis*, changing *ananas* – 'pineapple' – to *ananatis* – 'of the pineapple' – and so on.

Such matters of form may seem minutely trivial to the general public, but taxonomists can be jealous and argumentative creatures. These small changes elicited 'letters to the editor', complaining that the revisionists had shown a 'lack of respect for all those who use bacterial names in their professional life', as well as an appeal to the judicial commission that oversees such things, demanding that the (wrong) names be treated as *nomen conservandum* (see p. 136) – names that are accepted even though they do not comply with the rules. The appeal failed, if you are remotely interested. Nouns in apposition are generally frowned upon because they frequently result in errors and unseemly squabbles. It is a wild ride being a taxonomist.

Adjectives as generic names

A generic name can also be an adjective, in the form of a noun such as *Glaucus* (a genus of bluish green sea slugs) from the Greek *glaukos* for 'bluish green', or *Limulus*, from the Latin *limus* for 'askew or odd'. *Limulus polyphemus* is the horseshoe crab, which is distinctly odd all over, but particularly odd in appearing to have only one eye because its two eyes are very close together (hence *polyphemus* from the one-eyed giant described in the *Odyssey*). *Drosera* for 'dewy' and the plant genus *Angelica* for 'angelic' are two more examples of adjectives masquerading as nouns. They all imply the existence of a subject, so 'dewy one' or 'angelic one' will make sense of them.

Combining forms

I took my six-year-old daughter to the doctor some years ago. He asked 'What seems to be the matter?' I told him that Lily had a rash on the skin around her mouth. He examined her carefully and gave his considered opinion: 'She has perioral dermatitis.' When I pointed out that this was just Greek for 'skin rash around the mouth'* he laughed, but at least he had the decency to blush a little.

---

* In fact this is partly a 'bastard' term, with Greek and Latin combined: Greek *peri*: 'around'; Latin *oralis*: 'mouth'; Greek *derma*: 'skin'; Greek *itis*: 'disease'.

The name for the grey woodpecker is straightforward, but what about a woodpecker with grey-green plumage? As this example and Lily's rash indicate, it is combined forms that come to the rescue, providing nomenclaturists with endless opportunities for new coinages. The idea is not new. Linnaeus used or invented many names made from two or more words, such as *Scleranthus polycarpos* ('many-fruited tough-flower') and *Coccothraustes caerulea* ('blue kernel-cracker'). A more modern example is 1891's *Microdipodops megacephalus* ('thing that looks as though it has two small feet with a big head')* – the kangaroo mouse from North America.

These more complex names provide additional information about the species to which they are attached, although, as in the last example, they can be a bit of a puzzle. Despite sometimes having the qualities of the Enigma Code, this usage has afforded biologists endless opportunities to cram a great deal of information into the meagre two words of a binomial. These names have come almost full circle to the long, highly descriptive phrase names that were used before the binomial system was devised two hundred and fifty years ago. Now we have the tooth-billed bowerbird, *Scenopoeetus dentirostris* (*skēnē* meaning 'sheltered place' and *poiētēs* 'one who makes', plus *dens*, 'a tooth', and *rostris*, 'the beak'). There is also the tree mouse *Dendromus mesomelas*, which climbs trees and has a black stripe running down the middle of its back, clearly indicated by its name: *dendron* is 'tree' and *mus* 'mouse', plus *mesos*, meaning 'middle', and *melas*, 'black'.

Names such as *Microdipodops*, which combine three or more

---

* Briefly: *micro* ('small'), *di* ('two'), *pod* ('foot') and *ops* ('looks like'), plus *mega* ('big') and *cephalus* ('head').

words, are surprisingly uncommon. However, a few very long names do exist, such as *Notiocryptorrhynchus*, a beetle with a 'mark that is hidden on its nose' (Latin *nota*: 'mark'; Greek *crypto*: 'hidden'; Gk. *rhynchos*: 'nose'). Dinosaur specialists are among the most prolific authors of triple names, largely because they feel obliged to tack *saurus* onto the end of whatever else they have in mind. For example, there are *Micropachycephalosaurus*, a genus of 'lizard with a thick, tiny head' (Gk. *mikros*: 'small'; Gk. *pachys*: 'thick'; Gk. *kephalo*: 'head'; Gk. *sauros*: 'lizard'), and *Veterupristisaurus*, a genus of 'old, shark-like lizards' (L. *veterus*: 'old'; L. *pristis*: 'shark' or 'sawtoothed fish'; Gk. *sauros*: 'lizard'). Quoted in Theodore Savory's splendid little 1962 book on nomenclature we find *Polichinellobizarrocomicburlescomagicaraneus*. Unfortunately the original reference for this monstrosity has proved elusive, and apart from concluding that it may be some exceptionally weird-looking spider, we must consider it to be apocryphal. The longest binomen I can find is *Gammaracanthuskytodermogammarus loricatobaicalensis*, a species of sandhopper devised by the Polish entomologist Benedykt Dybowski in 1927. Dybowski was in his mid-nineties when he created this and several other notorious horrors, but it is more cheering to consider them the product of a playful old age rather than frailty of mind. I leave the etymology of this and the previous name for your own analysis as homework.

Greek/Latin hybrids

Quite which, and how, names can be compounded is governed by grammar, convention, good sense and good taste. One deadly sin is forming a new generic name or specific epithet from both Latin and Greek. Everyday English is awash with Greek/Latin hybrids, which all but the classically sensitive speak without a blush. 'Television', 'bigamy', 'neuroscience', 'homosexual' and 'sociology' are all offspring of an unnatural union between two separate languages, but whether one thinks of them as hideous aberrations or just useful words is entirely a matter of taste. There are no 'rules' to prevent these neologisms in common speech and none in biological nomenclature. But why get something wrong if you can so easily get it right?

The triple compound *Sinornithosaurus* derives from Latin *Sinae* for 'Chinese', Greek *ornis* for 'bird' and *sauros* for 'lizard'. This is not too bad, because each of the component words is too well established in nomenclatural usage to be easily replaced. But what was the American botanist Oliver Farwell thinking when he named a new genus of fern *Scyphofilix*, which compounds the Greek for 'cup' with the Latin for 'fern'? Mongrels can also be formed with words from non-classical languages, but they are very rare because most such words appear alone. It is worth mentioning that individual compound words are the issue here – there is no suggestion that the genus and species names within a binomial should come from the same language, because there would be no way of retaining consistency when a species moves to another genus. A Latin genus and Greek specific epithet is therefore perfectly acceptable, although it is most fortunate that the sixteenth-century botanist Pierre Richer de Belleval's bright idea of using Latin genera rendered in Roman characters and Greek specific epithets in Greek characters was never adopted. Many people find Latin binomials intimidating enough and would not take kindly to *Gentianella* ἐαρανθοκυανοζλωρος.

Often, compound words are just two words directly next to one another without a joining vowel. So the evergreen European box,

Buxus sempervirens ('always green' box), has nothing between *semper* and *virens*. This method is often applied to more or less amusing effect when proper nouns are conjoined. The palm genus *Roystonea* was named for the American Civil War general Roy Stone by O. F. Cook. Cook, in turn, was honoured with the name of the millipede genus *Ofcookogona* – the *gona* from the Greek *gonatos* for 'knee' or 'joint' (reflecting the obvious fact that millipedes are more joint than substance) joined to its component words with the vowel 'o'. The mushroom genus *Conocybe* is similar, being constructed from 'cone' and 'head' joined with an 'o' ('o' is the chosen vowel for Greek compounds, but had they been Latin an 'i' would have been used). However the components are joined together, it is important to get them the right way round. Hippopotamus, for example, has been criticised for not naming the creature 'river horse' as intended, but 'horse river', which is quite different.

Affixes

There is much more to compounds than just joining nouns or adjectives; nomenclature would be almost impossible without the debt it owes to prefixes and suffixes. Classicists sometimes complain that to the great unwashed words are just letters and sounds and that the complexity of derivation within even the commonest of words passes them by unregarded. I think this is unfair and largely untrue, but the way that affixes of all sorts are used to modify meanings of stem words is little understood or appreciated. Some stem words in English are so surrounded by prefixes and suffixes that they can be lost in the muddle. Bill Bryson in his engaging *Mother Tongue* points to 'incomprehensibility' as a prime example, with seven affixes and only 'hen' having independent meaning (from the Old English *cennan*: 'to make known').

Authors of Latin binomials love their prefixes and suffixes too, adding an extent of meaning unobtainable with naked adjectives and nouns. The botanist and writer William Stearn listed nearly eighty Latin and Greek prefixes and eighty common suffixes available to taxonomists, but these are just the standard ones that could be found in any Latin or Greek dictionary – there are hundreds more. Where would the nomenclaturist be without *-ensis*, 'comes from', to denote that the extinct 'hobbit', *Homo floresiensis*, was discovered on the Indonesian isle of Flores, or that the meadow waxcap mushroom, *Hygrocybe pratensis*, does indeed grow in meadows (*pratum*: 'meadow'). Or *semi-*, as in *Paneolus semiovatus*, the smooth-capped mushroom that looks like half an egg. *Semiovatus* is, in fact, nearly all prefix and suffix, the *atus* indicating 'likeness'.

Qualities of being under (*sub-* or *hypo-*), around (*peri-*), before (*ante-*), downwards (*de-*), good (*eu-*), big (*macro-*), small (*micro-*), similar (*-aceus*), capable (*-ellus*) and made-from (*-ineus*) can be conveyed. There are rules, of course. Suffixes need to agree grammatically, and different suffixes and prefixes must be used for nouns and adjectives. In addition, both suffixes and prefixes must be attached only to words that derive from the same language (words that derive from neither Greek nor Latin are treated as Latin). The nomenclaturist will always match a Greek affix to a Greek stem and a Latin affix to a Latin stem, hence the need for at least two affixes for every concept.

Suffixes

There can seem to be more suffixes (and prefixes) to go round than are strictly necessary, with many indicating the same idea; sometimes this is owing to the wealth inherent in classical languages, sometimes a matter of subtle differences in meaning. So *-ensis* is generally used for country or district of origin, but habitat may be served by *-cola*, as in *silvicola* (from woods). *Pepsicola*, incidentally, would be 'from the digestive tract', from the Greek *pepsis* for 'digestion', and I offer it as a helpful suggestion to anyone who names intestinal flora and fauna.

A more subtle idea is found in the suffix *-escens*, as in *Lactarius pubescens*, a mushroom commonly called the downy milkcap. *Pubescens* indicates that it is covered in fine, downy hair, because it means 'just beginning to get furry', *-escens* meaning that something is just beginning. If one wanted to suggest that a species displays an abundance or full development of some quality then *-ulentus* would do the trick, as in the common morel *Morchella esculenta* ('morel that is food at its best') or drought-tolerant *Succulentus arizonicus*, which has leaves full of juice (*succus*).

The summer grape from North America celebrates its fruiting time in the name *Vitis aestivalis*, using the suffix *-alis* ('of' or 'pertaining to'). Similarly, the mushroom *Galerina autumnalis* is 'of the autumn'. The genus name *Galerina*, based on *galerum* – 'bonnet' – becomes 'a little bonnet' with the addition of *-ina.*

The suffixes *-ibilis* and *-abilis* are immediately under-standable because they appear so frequently in English: 'ability', 'flexibility', 'vulnerability' and so on. Dangerously similar to the deadly *Galerina autumnalis* is the edible mushroom *Kuehneromyces mutabilis*. *Mutabilis* derives from *mutatus*, 'change', suggesting that it is 'able to change', reflecting its habit of changing from dark russet brown to buff from the edge of the

---

* 'Little autumn bonnet' is a charming name for something that will have you dead in short order.

cap inwards. Another fungal example is *Gymnopilus spectabilis*: 'naked cap able to appear magnificent'. *Variabilis* – 'a tendency to vary' – is an all-too-frequent specific epithet, which gives the impression that the taxonomist despaired of getting to grips with his type material. Incidentally, the suffix *-ibilis* has particular resonance among the British, thanks to the infamous Christmas Message of 1992 in which Queen Elizabeth referred to her family's *Annus horribilis*. The year had been difficult for many, and this was too obvious a Latin pun to pass unnoticed. A newspaper headline of the time summed up the mood nicely: '*What about our Annus, Ma'am?*'

Anus horribilis

This brings us to the most troublesome of all Latin suffixes: *-anus*. Having dealt with the embarrassments this has caused in the quarantined section devoted to such things (see p. 53), I will simply relate what it means and how it is used. Like most suffixes, it has several related meanings – 'pertaining to'; 'belonging to' – which overlap with other suffixes. The form *-anus* is masculine; there are also *-ana* (feminine) and *-anum* (neuter), which must agree with the noun to which they are joined. Familiar in everyday English, from 'Victoriana', 'Americana' and so on, the suffix has a near full-time job in biological nomenclature in honouring men and women of note.* Botanists are particularly fond of this usage, with the specific epithets *rivinianus* for the German botanist Augustus Rivinus (see p. 179), *raddeanus* for the German naturalist and explorer Gustav Radde and *wulfenianus* for Franz von Wulfen,

---

* There are other endings that perform the same function, the most common being *ii* and *i*, as in the (real!) epithets *smithii* and *jonesi*, and *a* as in *Sloanea*.

a botanist and zoologist from Belgrade, being a few of the many accidents waiting to happen.

An organism can belong to a country as well as to a person, hence the name of our possible ancestor and probable cousin *Australopithecus africanus*, as well as *Laminacauda fuegiana* (a spider from Tierra del Fuego) and *Viviparus georgianus* (a boring-looking snail from Georgia, US). However, the etymology of species names can defeat analysis, and all too often we are left with puzzles. *Solanum*, the genus that includes tomato, potato and aubergine, appears to mean 'belonging to the sun', an odd name for a genus that also contains the nightshades. Its true derivation is possibly from *solatium* ('solace') or *solor* ('soothe'), for its medicinal virtues, but even the reckless Culpeper suggested avoiding this treacherous genus.

More common suffixes

Specific epithets were first devised to provide differentia (see p. 150), so suffixes that convey the idea of difference in size or degree are frequently found. For 'small', there are *-cellus*, *-cillus*, *-culus*, *-ellus*, *-ullus*, *-illus* and *-iscus* (plus feminine and neuter versions, such as *-ella* and *-isca*). The mushroom genus *Campanella*, for example, means 'little bell', from *campanus* ('a bell'), while the chanterelle genus *Cantharellus* is 'little drinking cup'. Sometimes a little more thought is needed to tease out the meaning, as in *Oxalis acetosella* (wood sorrel), which means literally 'little-vinegar acid' or, more helpfully, 'acidic but not very sharp', which just about describes the pleasant taste.

Extremes or completeness can be indicated by *-issimus(a)*, as in the mushroom *Hygrocybe splendidissima* ('most splendid watery head') or the plant *Erigeron formosissimus* ('most beautiful fleabane'). One ending that causes confusion is *-aster* (plus *-astra* and *-astrum*). It means 'inferior' or 'incomplete', as in the wild olive, *Olea oleaster* – a poor thing compared with its cultivated cousin, *Olea europaea*. However, 'aster' etc can have a completely different meaning – 'star' – and context may be required to understand which meaning applies. For example, the extraterrestrial-looking fungus genus *Geastrum*, known as earthstars, is indeed star-shaped.

Fossil suffixes

The suffixes *-ites* and *-xylon* are unusual, because often they are tacked on the end of existing names to indicate the fact that the genera in question are fossils. For example, the familiar fungus genus *Agaricus* has a parallel existence in the fossil fungus genus *Agaricites*. While this is a convincingly focused name, the same can hardly be said for the desperately vague fossil fungus genus *Fungites* devised by Ernst Hallier in 1865.[*] Although *-ites* is used throughout the disciplines, *-xylon* finds a specialised use in fossilised wood. Again, names so adorned are usually nomenclatural extensions of existing genera, for example *Cupressinoxylon*.

The asylum of ignorance

Endings that suggest similarity are: *-oides*, *-aceous*, *-aneus*, *-ago*, *-atus* and *-icans*. The last of these means 'a tendency towards something', for example in *Boletus radicans*, which, most unusually, extends its stipe (stem) into the ground, almost exactly like a

---

[*] *Fungites* is certainly a fungus genus and a fossil genus. However, the species *Fungus fungites* is not a fossil species and not, rather startlingly, a fungus. It is a coral.

plant root. The first in the list is notorious, a suffix about which Linnaeus had nothing good to say. From the Greek *eidos* ('form'), *-oides* indicates, when attached to a noun, that something is similar in appearance to something else.

In a 1737 letter answering a question from Albrecht von Haller, 'Why does the termination *oides* displease?', Linnaeus wrote:

> *Because it is the asylum of ignorance. Botanists*
> *of the present day have scarcely introduced any*
> *new name but what ends in oides . . . I do not*
> *approve any names which differ only in their*
> *termination, or in having as it were different tails,*
> *or appendages . . . Every botanist who founds a new*
> *genus, should ascertain its essential character . . .*

He was concerned only with the names of genera, partly because his specific epithets had not yet taken full form. His objection was to existing genera having *-oides* bolted onto the end of them in order to provide a name for another, related genus; there is no reason why a comparison with something completely different would not have been acceptable. A fungus that looked like a coral – *Coralloides*, for example – would be both valid and useful. His opinion was that naming a genus should involve saying something positive about it, and that merely stating that it was similar to another genus just wouldn't do. The use of *-oides* suggests that the taxonomist could not think of any particular characteristic to include in a name, instead falling back on mere comparison with another genus.

Despite Linnaeus's objections, *-oides* has lived on regardless and is generally used sensibly by most. Not so well by ornithologists, though. They have a habit of doing just what Linnaeus warned against – adding -*oides* willy-nilly to the end of genera to make new ones: *Fringilla/Fringilloides*, *Geocichla/Geocichloides*, *Hemicircus/Hemicircoides* and many, many more.

Good usage is seen in *Hippoglossoides*, a genus of fish, which, one must presume, reminded the taxonomist of a horse's tongue. The *Arachnoides* are spider-like sea urchins, and the toadfish are in the genus *Batrachoides*, from the Greek *batrachos* for 'frog', which the author evidently considered to be near enough. What is to be made of the tiny sessile marine animals *Smittoidea* I do not know: perhaps Osburn, who named them, thought they looked like his predecessor, Herr Smitt. Even more puzzling is the Australian genus of beetles known as *Oides*, which appears to be entirely suffix.

The suffix -*oides* is more useful in specific epithets because it is not open to this type of abuse. It can qualify only the epithet, not the whole species name. For example, the field mushroom, *Agaricus campestris*, cannot have a lookalike called *Agaricus campestrioides* because it would mean that the mushroom looks like a field, not that it looks like *Agaricus campestris*.

The suffix is more popular in some disciplines than in others, botanists using it three times more often than zoologists. As with generic names, a common usage is to liken one organism to another. The name of the nocturnal raccoon dog, *Nyctereutes procyonoides*, fittingly means 'night runner that looks like a raccoon'. *Brachyteles arachnoides* is a slightly odd choice for the woolly spider monkey, because the genus name, meaning 'entirely short', is a reference to its short or missing thumbs. Thumbs or not, the whole thing looks a little like a spider. The fungus *Tricholoma inocyboides*, unusually for its genus, has the shape typical of another genus: *Inocybe*. This is spectacularly unhelpful to the layman, but the author clearly considered that anyone interested enough to look at a *Tricholoma inocyboides* would know what an *Inocybe* looks like.

Sometimes this does get a little out of hand, particularly, again, thanks to ornithologists. *Cygnopsis cygnoides* was once the name of the swan goose. The name means 'thing that looks like a swan that is swan-like'. The outright winner in the useless names competition is *Egretta egrettoides*: 'egret that looks like an egret'. At least it is not confusing, which is more than can be said for one unfortunate sedge. *Hypolytrum mapanioides* is a species that 'looks a bit like' a member of another sedge genus, *Mapania* (sensible enough). Another species was discovered within *Hypolytrum* that, evidently, 'looked a bit like' *H. mapanoides*. In a clear case of an affix too far, it was duly dubbed *H. pseudomapanioides*. *Pseudo* means 'false', but in a taxonomic context nearly always indicates that something 'looks a bit like' something else. So *pseudomapanioides* means 'something that looks like something that looks like the genus *Mapania*'.

A more sensible use for *-oides* is to compare the species with something that is not another organism, as in *spathoides*: 'spade-like'. Two species of fungi that I commonly encounter on my fungus forays (to the delight of students) are *Macrolepiota mastoidea* and *Amanita phalloides*.

Prefixes

Prefixes often indicate position: *ambi-*, *circum-* and *peri-* for 'around'; *contra-* for 'against'; *di-* and *dis-* for 'away from'; *intro-* for 'inside'; *prae-* for 'before' and so on. Size is commonly addressed with the prefixes *macro-* and *micro-*, for example in *Macrolepiota*, a genus of larger mushrooms distinguished from a genus of smaller but otherwise similar species, *Lepiota*.

Numerals make a frequent appearance in both generic and

specific names, reflecting the number of seeds, toes, petals, crests and so on that an organism might possess. The plant genus *Pentaglottis* (of the borage family), for example, has five tongue-like scales making up the corolla (petals), and *Spermophilus tridecem-lineatus* is the thirteen-lined ground squirrel from North America.

As observed above, *pseudo-* means 'false' but in nomenclature denotes similarity with another species or genus. For example, *Pseudohydnum* is a genus of jelly fungi that display thousands of little teeth on their lower surfaces, just like members of the genus *Hydnum* (hedgehog fungi). Mycologists are very fond of this construction, with *Pseudoclitocybe*, *Pseudoomphalina*, *Pseudolyo-phyllum* and many more.

The opposite of 'false' is, obviously, 'true', an idea expressed by the Greek prefix *eu-*. This receives many an outing in nomen-clature, occasionally fulfilling the same function as *pseudo-*, but, as it were, from the opposite direction. Ornithologists are especially enamoured of this odd revisionism, seeming to suggest that previ-ously named genera were not 'true' genera. So *Eudacnis*, *Eugallinago*, *Eugerygone* and *Euhyas* were derived from *Dacnis*, *Gallinago* and so on. Another example comes from a much higher, indeed the highest, taxonomic level. The Prokaryota (bacteria) are distinguished from the Eukaryota (everything else). As *kary-* is the Greek stem for nuts, the terms literally mean 'before nuts' and 'true nuts', a reference to the absence or presence of a nucleus.

Eu- also means 'good', 'well', 'nice' and 'typical'.* *Eucalyptus*

---

* Not every *eu* has been added as a prefix by a taxonomist: some come directly from proper nouns or from existing Greek words. There is *Euphorbia*, after the Roman Euphorbius; *Euonymus* after Euonyme, the mother of the Furies, a lady 'of good name'; *Eupatorium* for King Eupator, a 'good father'; *Eucharis*, 'full of grace', and so on and on.

translates as 'well covered' from the Greek *calypha*, a reference to the calyx, which covers the flowers like a lid. Similarly, the penguin genus *Eudyptes* is a 'good diver'. Sometimes, however, taxonomists don't seem to be doing their job properly. *Euchoreutes naso*, the long-eared jerboa, is a 'good dancer with a nose'. The presence of a nose is, frankly, an unremarkable fact, and it is an unremarkable nose, but how is it that the nomenclaturist failed to mention either of this creature's two extraordinary and outstanding features? *E. naso* has a tail twice the length of its body and with a conspicuous 'duster' at the end. Then there are its ears: huge appendages that would have Dumbo green with envy. It must be suspected that Philip Lutley Sclater, the great nineteenth-century zoologist who named it, had a wry sense of humour.

In 1998, the palaeontologist Jennifer Clack described a fossil to which she gave the name *Eucritta melanolimnetes* ('true creature from the black lagoon'). I have a suspicion that the prefix is merely an attempt to provide a classical gloss on the decidedly unclassical 'critta', however, I wrote to Professor Clack, and she kindly told me that the prefix merely expressed her view that the creature was beautiful.

If the cap of a mushroom or the leaves of a plant were covered in flattened hairs, it might be described as *tomentose* (Latin *tomentum* means 'pillow stuffing'). If they were sparse, it would be *subtomentose*. An organism that was blue, but not very blue, might be *subcaesius*. The prefix *sub-* means 'nearly' or 'just below', and it is used, perhaps regrettably, to say that something is like something else. For example, the flycatcher genus *Sublegatus* is 'nearly *Legatus*', another flycatcher genus. The most famous *sub-* is *subbuteo*. The table-top football game was invented by the Kentish man Peter Adolph, just after he had spent his honourable war years in the RAF no doubt apologising for his surname. His attempt to register the game under the name 'Hobby' was thwarted by the patents office, which thought it too generic, so he went instead for the Latin name for the hobby, *Falco subbuteo*: 'falcon that is almost a buzzard' (from the Latin for buzzard, *buteo*).

Naming hybrids

The matter of naming hybrids is quite awful, laying confusion upon complication. Hybrids are rare in zoology (although not as rare as once thought[27]), so hybrid animals are viewed merely as individuals with no nomenclatural status. If matters were quite so simple among plants, botanists would have been spared the thirty-one articles on how to name hybrids that have been published in the *International Code of Nomenclature for algae, fungi, and plants* (ICN) (see p. 102). Hybrids can occur between plant species belonging to different genera or between plant species within a single genus. Provided it breeds reasonably true, there is no particular reason why the name of a genus or species should reflect its shameful parentage, so many pass unnoticed. *Digitalis mertonensis*, for example, is a hybrid between two other species of *Digitalis* (foxglove).[28] However, some indication of hybrid status is usually given. A common method of denoting a hybrid is the use of '×', and *Digitalis* × *mertonensis* is the correct rendition. Similarly the willow species *Salix lapponum* and *Salix reticulata* can hybridise to form *Salix* × *boydii*.

For a hybrid between two or more species of different genera, the '×' is placed before a newly coined generic name, different from the parental generic names and usually made up of parts of each. So, by combining all or part of each name (but not all of both names!), ×*Agrohordeum* is a hybrid genus deriving from the crossing of two species of grass belonging to *Agropyron* and *Hordeum*. Given the ban on reproducing both names in full, ×*Ericalluna* from the heather genera *Calluna* and *Erica* is an impressive workaround. David Gledhill in *The Names of Plants* notes the splendid hybrid orchid genus name derived from a dalliance between *Gastrochilus bellinus* and *Doritis pulcherrima*. It is, of course, ×*Gastritis*.*

Plants can be serial hybridisers, resulting in multiple parentage. Up to three† parental generic names can be used to form a new name, but for more than this (often for just three) a completely new name is devised, traditionally based on the name of a collector or grower and adorned with the special suffix -*ara*. Species from the three orchid genera *Arachnis*, *Renanthera* and *Vanda* cross-bred to form a hybrid species in a hybrid genus newly designated to accommodate it: ×*Cogniauxara*, for the Belgian orchid specialist Alfred Cogniaux.‡

An alternative is to place the special prefix *notho-* before a generic name, or indeed before a specific name, for example in *Nothopanax* and *Nothobaccharis*. The ICN coyly translates *notho* from the Greek as 'hybrid', but a straightforward dictionary definition renders it more plainly. In both Greek and Latin, it is 'bastard'.

---

* In all cases the '×' is not considered to be part of the name, just an indication of status; this is reflected in the fact that it is never rendered in italics.

† Having three or more parents does not require the services of a mad scientist – an existing hybrid breeds with another species from another genus.

‡ Orchids have presented sufficient difficulties in naming hybrids to require their own code, the *Handbook on Orchid Nomenclature and Registration*. There are more species of orchid than of any other plant genus (more than twenty thousand and counting) and they are world-class hybridisers. The International Orchid Register is an online database where 'species' are not searched for by their name but by the names of their parents.

Ballocephala verrucospora

CHAPTER IV

The ACT of NAMING

THE BRITISH MYCOLOGICAL SOCIETY AUTUMN FORAY IS A WEEK-LONG meeting of charming, erudite and sometimes slightly eccentric people with a common purpose: studying fungi. Crates of books, computers, microscopes, drying equipment and chemistry sets are unloaded from the boots of cars and laid out on benches in a field-study centre or university laboratory. The average day entails a field trip or two, followed by time in the lab identifying the finds of the morning, then dinner, then pub.

On the first autumn foray I attended, in Lincolnshire, I noticed one of our company carrying a shoulder bag instead of the traditional basket. Inside was a collection of pots. The man showed little interest in the forest floor or the fungi that grew from tree trunks, but became very excited when we came across a small, bubbling stream. Taking one of the pots out of his bag, he collected some of the frothy scum that had accumulated in a little stretch of 'rapids'. He had been fairly taciturn up to that point, but when I asked him what he was collecting, there was no stopping him.

His passion, I found out, was the 'aquatic hyphomycetes', and he was collecting their 'conidia' (asexual spores). These, he told me, far from being uninteresting (my expression was evidently inappropriate), were fascinating, important and beautiful. They were, he said, exquisite structures, some shaped like stars, others resembling shrimps or snails or trees or snakes. I learned that these microscopic spores could be found in vast numbers in the froth of streams and that they were responsible for the breakdown of leaves that fall into the water.

I have met hundreds of people like my aquatic hyphomycete enthusiast over the years. There was Peter, who studied fungal gnats (known to mushroom hunters as the maggots that crawl out into their frying pans); Phil, who found micro moths so much more interesting than butterflies; Alan, who became so delightfully obsessed with *Boletaceae* (the mushroom family that includes penny buns) that he built an extension to his house as a laboratory. Stewart spends months at a time looking for insectivorous plants in Indonesia, while Dr Morris knows more about weevils than any man should. I could go on. Some of these biologists are professionals, attached to a university or botanic garden. Others are amateur naturalists pursuing their private passions. All have the charm of the true enthusiast. These are the men and women who devise the scientific names by which the objects of their obsessions are known.

A *species of one's own*

Mike Richardson is another such enthusiast. A fellow devotee of fungi, he and I were sharing an Indian meal in an Edinburgh restaurant one evening when he told me the story of *Ballocephala verrucospora*. Back in the late sixties, he was attending his son's first birthday party, something which, frankly, can be a bit of a trial for the adults tasked with keeping the little darlings happy. Unable to face a gruelling afternoon of cake and tears, he decided to go for a walk instead. It took him to the Pentland Hills near

Edinburgh, where he had only sheep for company. Where there are sheep there are sheep droppings, and knowing that sheep droppings can be more interesting than most people might suspect, he took some home to incubate.

A few days later, Mike noticed some small white clumps on the surface of the droppings. Studying his treasures with the help of a dissecting microscope, he discovered that these white clumps were growing from the surface of deceased tardigrades. A tardigrade (its name means 'slow walker') is a tiny creature (1.5 millimetres long makes one a giant of the clan), with eight short, fat legs and an ambling gait that has provided it with the common name of 'water bear'– the 'water' part coming from its preference for damp habitats. Mike likes tardigrades (who wouldn't?), but it was the white clumps that interested him because they were a fungus. Many small creatures can end up as a fungus's dinner, consumed by the thread-like hyphae that invade their bodies and suck out their precious bodily fluids. A search of the literature enabled Mike to assign the fungus to an existing genus, *Ballocephala*. But he noticed that its spores were larger than those of the only species known at that time, and were covered in little warts. This and other considerations led Mike to believe at first that he had an undescribed variety of *Ballocephala sphaerospora* on his hands. However, a fellow mycologist was sure that the find was sufficiently distinct to constitute a new species.

Mike's publication of his new species followed the typical pattern that had been used for more than a century (although such a publication is likely to be considerably more complex nowadays than when Mike was writing back in 1970). His paper (referred to as a 'protologue' because it is the first word on a species*) begins with an introduction, explaining why he considers his find to be a new species. Then comes the description proper, starting with the heading: 'Ballocephala verrucospora sp.nov.'

Judith Winston in *Describing Species* wrote that while some people approach the naming of a new species with trepidation, others are like expectant parents, asking friends for suggestions and making lists of good candidates. *Ballocephala* was the name already given to the genus by Charles Drechsler in 1951. It means 'head thrower', a reference to the way in which the packet containing the spores (the sporangium) is released by being thrown by the arm-like sporangiophore. The specific epithet *verrucospora* was provided by Mike himself and simply means 'warty spores'. This was a good choice, because it describes one of the characters that 'differentiates' this species from *B. sphaerospora*, although there is no rule to say that a specific epithet must represent a difference (differentia in the trade, see p. 150).

The abbreviation 'sp. nov.' stand for *species nova* or 'new species'. Because it is a new species there is no reference to previous descriptions or authors, as would have been necessary had he been reviewing an existing species. In a nice piece of conventional modesty, now dispensed with, Mike does not cite his own name – that is for later writers to do.

The next part of the description begins: 'Hyphae hyalinae, inclusae in corpore hospitis, constantes de cellis secedentibus 20-40 x 10-12 μm diam. Sporangiophora crescentia per superficiem dorsi hospitalis, 50-150 μm alta x 5-7 μm diam, septata solum ad basem . . .' This is the beginning of the Latin description, which is one hundred

---

* Linnaeus set the template for the protologue.

words in this case but could be shorter or longer. For a mycologist, even one whose Latin is a hazy memory from school, this description is not too difficult to understand. The Latin of biology is quite different from that of Ancient Rome and different again from Church Latin. It is a highly stylised language all of its own, brimming with specialist Latinised terms that would make Mike's description almost completely incomprehensible to Cicero. *Hyphae hyalinae* means something like 'glassy web', but here a mycologist would understand it to mean that the hollow fibres (hyphae) that are the 'cells' of the fungus are transparent (hyaline). All branches of biology have developed their own specialised nouns and adjectives, usually derived from Greek or Latin, so a description written by a zoologist might make little sense to a mycologist or botanist.

As is traditional, the description is repeated in English translation: 'Hyphae hyaline, internal in the body of the host, composed of disjointed cells 20–40 x 10–12 μm diam. Sporangiophores growing through the dorsal surface of the host, 50–150 μm high x 5–7 μm diam, septate only at the base.'

What Mike provides is a 'description', but he could have given us a description and something called a 'diagnosis', or even just a diagnosis. Diagnoses explain what is sufficiently different about the species in question to qualify it for species status. They can be extremely short. For example, if all that differentiated a new species of rabbit from others in its genus was that it was blue, then 'blue', would be a sufficient diagnosis. A diagnosis has long been required in zoology but is optional in botany and mycology, where either or both are sufficient.*

---

* Mycological journals now usually insist on a short diagnosis followed by a description, something that is likely to be required for all publication of species under the code for botanical nomenclature.

Below the description comes this line: 'Habitat parasitus in Tardigradis (Macrobiotus?). In fimo ovino, West Kip (550 m), Midlothian, Scotland, II. i. 1969. Typus IMI 148042.' This tells us the habitat – parasitic on tardigrades; that the tardigrade could possibly be a member of the genus *Macrobiotus*, and that it was found in sheep's dung ('*In fimo ovino*'). It also states where it was found. '*Typus IMI 148042*' is very important. It tells anyone who needs to check on the original specimens studied by Mike where they can find them – in this case at the International Mycological Institute, reference number 148042. More on 'types' later (see p. 115). Just before 'Typus' is '11. i. 1969'. This is the date on which the specimens were discovered and also, of course, someone's birthday.

The paper includes some line drawings and scanning electron micrographs (which were pretty cool things to have in those days) and ends with further descriptive material and discussion. It was published in *Transactions of the British Mycological Society* after the normal peer review. Like all scientific papers, its content would have been judged by the scientific community. If anyone had disagreed with Mike's assessment, they would have presented their own paper to *Transactions* or some other journal. Whether or not the species name and description fared well subsequent to publication, the species will always be referred to as '*Ballocephala verrucospora* M. J. Richardson'. It is still accepted as correct on the many databases that record these things, and is occasionally referenced in other papers, so it is a 'good' species. There is no authoritative body that 'accepts' a species; either it is accepted by the interested scientific community or it is not. However, taxonomists do not always agree, and the acceptance of some species can be argued about for decades.

There have been many innovations in taxonomic literature since Mike's paper was published, but one, effective from 1 January 2012, stands out as, if not revolutionary, then at least a major break with the past. The international code that covers plants, fungi and algae no longer requires a Latin diagnosis or description for valid publication – they can be written in Latin or English. Latin was customary throughout most of the history of nomenclature and became mandatory in botanical nomenclature in 1935, so this is a big step, which has been widely applauded – notably by the Chinese, who were thrilled (although I do not know what the French thought of the matter). The universal nature of Latin as an academic language has been finally and inevitably usurped by English. Zoology is way ahead of botany in that it has never insisted on Latin, except in the names themselves.

Unlike other scientific papers, a published taxonomic paper with a valid description of something new cannot be forgotten. It is a 'nomenclatural act', and its contents must be taken into account in further publications that touch on the same taxonomic group. Most particularly, a species name, once established, cannot be used again, even if the species it applies to turns out to have been a figment of the taxonomist's imagination (as has happened, see p. 47).

The term 'validly published' is an important one. Most people will think of it as meaning a peer-reviewed taxonomic paper written in accordance with the rules and published in a printed learned journal or book. But now online publication alone, in Portable Document Format (the familiar PDF file), is considered sufficient for the paper to be validly published. Nowhere do the rules say that a paper must be published in a learned journal, online or in print – just that multiple copies should be readily available. And if 'learned' is dropped, then so is 'peer review'.

In 1984 and 1985, a whirlwind swept through the taxonomy of Australasian reptiles with the publication of three papers by two herpetological enthusiasts, Richard Wells and Ross Wellington. They had been unhappy with what they considered to be the moribund state of herpetology in Australia and New Zealand, and their papers were tossed into academia like firecrackers into a hen house. The papers, self-published in their own journal, the *Australasian Journal of Herpetology*, named new species and genera, renamed old ones and resurrected names that had previously been discarded. In total, they made 739 taxonomic changes. The professional scientific community was not pleased: ninety-one authors petitioned the body that controls zoological nomenclature (ICZN, see p. 101) to have the papers suppressed. The ICZN dismissed the petition, pointing out that the papers were validly published in accordance (mostly) with the rules. Just because no one liked the changes did not make them invalid in the nomenclatural sense. Even if everything in their papers had been scientific nonsense, which it wasn't, the nomenclatural acts themselves would still be valid – all 739 of them.

Mike's considerably less controversial paper was written nearly half a century ago, obeying the conventions and rules for naming as they were at the time. However, times change, and modern scientific papers can be much more substantial affairs. They follow a fairly standard format of Title and authors, Abstract, Introduction, Methods and materials, Results, Conclusion, Acknowledgements, References. Papers naming new species usually have an Etymology section, explaining why the name was chosen (at least they do now – there are many mysterious, unexplained coinages from the past).

Within this overall structure, taxonomic papers vary enormously according to what sort of taxonomy they are doing, how many taxa they are doing it to and the conventions of the publishing journal. If the paper is describing a new species, the description will usually be in the Results section (although it may be renamed 'Description'). Many papers are written that introduce a new species or two within a paper that re-describes all the species of a particular genus or higher taxon – this is usually called a 'systematic account'. These are, of course, easy to spot, with their 'sp. nov.' suffix. However, descriptions are not always of new species. They can be new descriptions of already described species (perhaps written in the light of new evidence) or descriptions of subspecies. Monographs, which give a comprehensive account of an entire taxonomic group, can also contain new and revised species descriptions. These are sometimes published separately as books, and may represent the life's work of an author.

Systematic accounts, or 'revisions', as they are often called, might move organisms around in the taxonomic hierarchy, rename old species, name new species while unceremoniously dispensing with others and generally ruin the preconceptions of all but the true specialist. As such they can be received with less than enthusiasm by the ordinary naturalist, because they mean that everything you thought you knew about a family or a genus was wrong, and that you will have to forget a great many names and learn many new ones.[*]

---

[*] A typical systematic account is that of the microscopic fungus genus *Plagiostoma* (L. C. Mejía et al 2011). Its authors set out their extensive aims, one of which is stated as follows: 'In this paper these eight new species of *Plagiostoma* are described, four species are redescribed, and four new combinations are proposed.'

One would think that writing a taxonomic paper would be straightforward enough for a specialist in the field, and that the establishment of a new name would automatically fit neatly into the body of knowledge within a particular kingdom. However, there is a very great deal that can, and has, gone wrong, so rules are needed to keep everything in order.

Prior to the systematising work of Linnaeus, there was taxonomic chaos, with a huge proliferation of names and methods of naming and classification. Although Linnaeus was a stabilising influence, it cannot be said that his efforts were entirely successful. New species continued to flood in from newly explored areas of the world and previously unsuspected species were discovered through the lenses of ever-improving microscopes. Naturalists went on sprees inventing new (binomial) names for species that they thought Linnaeus had missed, or changing existing binomials to names that were more to their liking. Synonyms (several names for one species) and homonyms (several species for one name) proliferated.

Attempts were made to introduce rules to bring some order to the naming of nature. In the spirit of the endless disagreement that has remained a constant throughout taxonomic history, the first system, the Strickland Code of 1842, was devised for zoologists alone. It was thought that the rules laid down by Linnaeus and others, such as the Swiss botanist Augustin Pyramus de Candolle, had served botany well enough and that zoologists were more in need of a code. Anyway, there was a minor tiff between botanists and zoologists concerning how authors' names should be recorded with species names – a matter on which they still cannot see eye to eye (see p. 113). A variety of codes were attempted after that, with separate codes for birds, insects, fossils, animals that weren't birds or insects, and plants, threatening a worse chaos than the one they were designed to avoid.

At the first International Botanical Congress in Paris in 1867, a botanical code, the Paris Code, which contained most of the important rules that still govern botanical taxonomy today, was agreed on a trial basis. After forty years it seemed to be going well enough, so in 1906 it was confirmed, with some important additions, in Vienna to form the Vienna Rules. Even then there was not unanimity: many American botanists did not much like what was agreed and used their own code – the American Code. To make matters worse, the Americans were not of one mind: two important books on the flora of the north-eastern United States, for example, followed different rules, often resulting in different names for the same plant.[29] Fortunately this split was eventually healed when the *International Code of Botanical Nomenclature* (ICBN) was established as the Stockholm Code in 1952.

There is also a code for the nomenclature of cultivated plants, the International Code of Nomenclature for Cultivated Plants, which allows and manages names such as *Acacia terminalis* 'Tasmanian Pink' and *Acacia* 'Gold Lace'. Zoology went through a similarly exhausting process, eventually arriving at the *International Code of Zoological Nomenclature* (ICZN) in 1958.

Bacteriologists, meanwhile, wanted their own code. To them, animals, plants, fungi and algae are Johnny-come-latelys, of little interest except as a source of food and shelter for the objects of their passion. They are quite right, considering that the living world is more or less divided up into prokaryotes (bacteria and allies) and eukaryotes (everything else). Bacteria had been included within the ICBN, but this code insists on type material that is thoroughly dead – something not possible with bacterial types (samples), which can only be preserved in living culture. The forerunner to the *International Code of Nomenclature of Bacteria* (ICNB) was first published in 1958; it covered bacteria and viruses, although not some of the cyanobacteria, which are still governed by the 'botanical' code for the unconvincing reason that they were long thought to be algae (blue-green algae). Virologists, who (again correctly) thought their domain far removed from the rest of the living world, duly set up their own code shortly after. Taxonomists spend their working lives dividing up the living world, but seem to spend their spare time dividing up themselves.

A long overdue change took place in 2011, when the botanical code (ICBN) was renamed. Mycologists had quietly seethed with resentment for years at following a code of botanical nomenclature. Botany is about plants, and fungi are not plants, not even a little bit. No doubt phycologists (algae specialists) felt the same, although they have long been a wandering tribe and may not care so much. It is a tribute to the amiable and reasonable temperament of mycologists that a compromise was reached and another messy divorce avoided. The proud new name, which was settled upon at the Melbourne meeting (with the three specialties placed in diplomatic alphabetical order), is the *International Code of Nomenclature for algae, fungi, and plants* (ICN).

There have been a few problems with species that everyone thought were animals, but weren't, and species that everyone thought weren't animals, but were. (Some zoologists have spent their lives telling people at parties that they are zoologists, only to find out that they have been mycologists all along). There are many organisms that have never been sure what they are and thus which code they should obey – so-called 'ambiregnal' organisms. Should they ever make up their minds, there is a chance that they may find they have been obeying the wrong set of rules, and chaos may ensue. The various codes inherit an early and naïve understanding of a natural world consisting of just plants and animals and, maybe, the larger fungi. But these three groups are now understood to be merely highly visible (i.e. macroscopic) organisms situated on a few of the many branches of the tree of life.

To have half a dozen codes is inconsistent and unwieldy, and an attempt to unify nomenclature is being made in the form of the BioCode. This would eventually standardise the rules across the living world, giving ambiregnal species an automatic home and removing all the confusion created by having several sets of rules. The BioCode continues to progress, with the latest draft being published in 2011. However, with representatives of the major codes sitting on the BioCode committee, its aim is to slowly guide the existing codes closer together. When sufficiently close, they will be unified in what would by then be the Biocode.

A new code, the Phylocode (see p. 263), threatens to upset apple carts because it takes the evolutionary history of an organism extremely seriously and uses it as the very basis for its system. Being more concerned with taxonomy (how things are arranged) than nomenclature (how they are named), the Phylocode sits uncomfortably with existing codes and, to put it as politely as I can, has not enjoyed a great deal of popularity within the scientific community.

c∞ɔ

The various international committees hold meetings at exciting locations every few years to review the rules and issue new rule books. The ICN, for example, is a monstrous tome, written in legalese compounded with botanese. I do not criticise at all – it has to be like this to allow the wheels of taxonomy to run, if not smoothly, then at least in the right direction.* As a system of rules developed during one and a half centuries, it has seen everything and knows all the pitfalls and anomalies that can arise.

The large number of rules and recommendations govern such esoterica as orthography (spelling and typography) and the acceptability of names, and ensure that only one correct name is applied to any one organism (or other taxon) within any circumscription.† The committees also decide which names can be conserved (retained) despite not quite fitting in with the rules, what constitutes the valid publication of a name and a thousand other matters. But it is important to note that these organisations do what their labels say they do – provide codes or rules for nomenclature. They do not (as they go to some pains to stress) take any particular view on taxonomy – evolutionary, *scala naturae*, alphabetical order, it is all the same to them.

Nor, despite their international standing, undoubted authority and legalistic turn of phrase, do these rules have any general legal standing. Anyone can call a species anything they

---

* It is translated into reasonably understandable English by Nicholas Turland in *The Code Decoded: A User's Guide to the International Code of Nomenclature for Algae, Fungi, and Plants.*

† A circumscription is a view on the placement of a taxon. For example, botanists disagreeing over which genera belong in the plant family Rubiaceae and which in the plant family Spermacoceae have different views about the circumscription of these families. Circumscriptions apply throughout the taxonomic rankings.

like and describe it in whatever way they like – it is just that no one else is likely to use it. However, as the Australian herpetologists demonstrated, it is perfectly possible to name species without access to established learned journals. This provides an opportunity to seriously annoy any scientist on your 'list'.* With a bit of effort you can produce endless names for imaginary species and completely mess up their entire specialty.

---

* The four people at the top of my 'list' are all mycologists.

Gorilla gorilla gorilla

CHAPTER V

The RULES of NAMING

THE MAIN PURPOSE OF NOMENCLATURAL CODES IS TO PROVIDE A single, stable name for each species. Unfortunately, 'single' and 'stable' are not easily obtainable objectives. What happens if a species acquires a second name when someone who is unaware that it has been already named devises a new one? Which name is retained? The same name may be used twice for different related species, or a name may be revised in some way by another taxonomist. How are such matters resolved? Nomenclatural codes supply rules to govern all these situations, as well as guidance on how names should be constructed and written down. Many can be traced back to Linnaeus's *Philosophia Botanica* (see p. 197), but time and convention has expanded them.

A scientific name must consist of two parts: the generic name and the specific epithet. Zoologists call these two-part names 'binomens', while botanists, phycologists (algae specialists) and mycologists call them 'binomials'. Subspecies require an extra name, making three, as in the white-cheeked gibbon, *Hylobates concolor leucogenys*. The latter should be rendered as *Hylobates concolor* ssp. *leucogenys* or *Hylobates concolor* subsp. *leucogenys* in formal writing. These are the zoologists' 'trinomens' and the botanists' 'trinomials'.

Zoologists do not formally recognise any rank below subspecies, but botanists and other disciplines governed by the ICN certainly do, and seem inordinately fond of them. In fact, there are four hair-splitting levels below subspecies – variety (var.), subvariety (subvar.), form (f.) and subform (subf.). Unlike higher taxa they are independent of one another. For example, it is possible to have a form without it being nested within a subspecies or variety. There is a suspicion that the use of such subtle distinctions reveals a creeping uncertainty, and it is true that it takes a particular type of mind to decide whether something is a subvariety or a form. However, from my own experience in mycology, I know of several clear species that come in a variety of colours, and these certainly deserve nomenclatural recognition. A typical example is *Inocybe geophylla* var. *lilacina*, the commonly occurring lilac variety of an otherwise white mushroom.

Linnaeus permitted almost any word to be used as a specific epithet (see p. 223). Modern rules and usage have extended this laissez-faire attitude to the genus. For example, the zoological code states that 'a name [note: generic or specific] may be a word in or derived from Latin, Greek or any other language (even one with no alphabet), or be formed from such a word. It may be an arbitrary combination of letters providing this is formed to be used as a word'.

This is liberating stuff, providing an extraordinary freedom for the taxonomist. However, the various codes do impose bewilderingly complex rules that govern the structure and spelling of names. This complexity is compounded by the fact that the codes are not entirely at one. However they generally agree that the name must be rendered as though it were Latin, excepting that all twenty-six letters of the English alphabet are available (the Latin alphabet is missing j, u and w). They should be grammatically correct, should consist of generic and specific names that are both short but at least two letters long, and (believe it or not) they should be pronounceable.

Letters from other languages have to be transliterated, so, for example, *mjøbergi* becomes *mjobergi*. Apostrophes are also unwelcome: *d'urvillei* is rendered *durvillei*. If either the generic or specific name is made up of two or more words, they must be joined together to form one word, sometimes by using a hyphen. The botanical code is not terribly keen on hyphens, but tolerates them when used by the original author and the component words are distinct, as in *Auricularia auricula-judae* (*auricula* means 'ear' and *judae* 'Jew', hence Jew's-ear fungus). The zoological code likes hyphens even less, allowing them only in rare circumstances, such as in *Polygonia c-album* (comma butterfly), which has a small but very noticeable white 'c' marking on the underside of its wings.

The zoological code sums up the tenor of suitable names for all, saying that as far as possible they should be 'appropriate, compact, euphonious, memorable, and not likely to cause offence'. Much beyond this commendable unanimity they go their separate ways. For example, the ICN fiercely rejects tautonyms, accepts diacritic marks, such as ä, é and so on, likes hyphens (if not very much), accepts hybrids and forbids generic names that refer to morphological characters.* The ICZN seems to adore tautonyms, hate diacritic marks (so ä becomes ae), accept hyphens only under duress, refuse to have anything much to do

---

* This rule does not apply to names published prior to 1912, which is why we still have the truffle genus *Tuber* ('swelling or growth') and the plant genus *Radicula* ('root'), which were published earlier.

with hybrids and does not appear to care about generic names with a morphological flavour.

Italicisation

The italicisation of scientific names is a matter on which naturalists have very strong views. Despite this, nowhere in the ICN is there a rule or even an official recommendation that names should be printed in italics. Indeed, the code explicitly states that it 'sets no binding standard in this respect, as typography is a matter of editorial style and tradition, not of nomenclature'. However, in the preface to the code, it is suggested that to do so is good practice, and the code sets an example by doing just that in its own usage.* Similarly the ICZN has no rule, but it does recommend as good practice the rendering of scientific names in a different font or format. Thus *Sorex araneus*, <u>Sorex araneus</u>, **Sorex araneus** or Sorex araneus are all acceptable, although the first is certainly to be preferred in most situations.

The practice of italicisation stems from the typographical convention that foreign words should be treated this way. For example: '*Entre nous*, the *objets d'art* were arranged in a rather *ad hoc* manner.' As this abomination of a sentence shows, it is easy to sound embarrassingly pretentious, and for this reason voices have been raised in objection to the use of italics in the serious work of biology. In addition, Latin names are daunting enough, it is said, without the addition of italics to indicate that the writer thinks he is cleverer than the reader. But conventions are hard to overthrow, and a Latin name printed without

---

* In the standard work on British plants, *New Flora of the British Isles*, Clive Stace uses italics only within the text; the species names used in headings are in bold and not italicised.

the italics will usually assure the reader that he is cleverer than the writer.

With this in mind, I always painstakingly italicise every Latin name I commit to print, even though it is tempting to leave a fiendish trap for pedantic and vocal critics by not doing so. In a series of articles I once did for a newspaper, I was mortified to find that my sub-editor consistently, and presumably painstakingly, removed every italic in sight, even when I begged him not to. Italics are here to stay and I do not think it a bad thing, because they make often-difficult words stand out a little more.

Capitalisation

Two typographical rules that do apply to the writing of a binomial are that the generic name should be capitalised and the specific name rendered entirely in the lower case: *Sorex araneus*, never *sorex araneus* or *Sorex Araneus*, and certainly not *sorex Araneus*. Although specific epithets are not now capitalised, in earlier names they often were, notably when derived from a proper name, as in *Berberis Darwinii* – Darwin's barberry. Anyone refer-encing it now would probably replace the capital D with d, although there is no rule to say that these historic names must be converted. Conventionally the generic name is shortened to its capitalised initial when the full name has been provided earlier in the text and there is no possibility of confusion. So *Sorex araneus* becomes *S. araneus* in subsequent mentions. It is very bad form to spring on a reader the un-introduced initial letter of a genus on the assumption that he or she will know what you are talking about. *A. domesticus*, for example, could be a mosquito (*Aedes*), a cricket (*Acheta*), a spider (*Araneus*) or a mushroom (*Agaricus*).

The one place where a writer can be free from all typographical constraints is in the heading section of a taxonomic paper. Here he can use anything he likes (or at least anything that the publishing journal will tolerate), so SOREX ARANEUS may be perfectly acceptable.

Citations

To be a useful name, a binomial preferably should be followed by the name of the author (or *auctor*, as botanists refer to them). Biological nomenclature is plagued by something called homonymy, where the same name (for a genus or a single species) makes two or more appearances. If you know the author's name, you know which *Tapeinanthus* or which *Astragalus rhizanthus* is being referred to. If you are wondering how a name acquires two or more authors, it is because taxa are sometimes redescribed or reassigned by later taxonomists and occasionally through error.

Author citations can also supply a pocket history of a name, enabling the reader to track its nomenclatural progress through the years. This adds another (essential) layer of complication to the naming system, albeit one that is seldom employed or needed outside of academic publications. The simplest author citation would appear as '*Agaricus campestris* L.' (the field mushroom, which was named by Linnaeus). This is not too onerous an addition to a binomial, but because species increasingly appear to be named by committee, their names – which are often long enough anyway – can acquire a not inconsiderable extra burden: *Alkanna mughlae* H. Duman, Güner & H. Şağban, for example.*

---

* In fact, beyond two authors, the name of the first author plus 'et al' is usually employed.

A more serious problem occurs when a species is named in a paper-within-a-paper. For example, a paper by Bean, Gawn and Dunnit on the ecology of salt mines might include the description of a new species of vinegar fly by Dunnit and Glad, making the species description difficult to find.

As with everything else in nomenclature, citations are subject to rules. There is nothing that could go wrong with *Agaricus campestris* L. – provided that one resists the temptation to put a comma after *campestris*. But few things in biological nomenclature are simple, and the issue of author citations is one of baroque complexity. What if a taxonomist named a species but there was something wrong with the way it was published ('invalid' in the jargon) and the species was subsequently re-described by another taxonomist using the same species name? What if the species moved from one genus to another, keeping its specific epithet but (obviously) losing its original generic name? How should the author names be written and in what order if there were more than one? And so on.

The two main nomenclatural codes differ in their rules and recommendations. Under the ICN the author name can be abbreviated; under the ICZN it cannot. For example, field mushroom is written '*Agaricus campestris* L.'. The ICN has a list of standard authors' abbreviations: L. for Linnaeus; Fr. for Elias Fries; Juss. for Antoine Laurent de Jussieu; R. Br. for Robert Brown and so on. But the rhinoceros is presented as '*Rhinoceros unicornis* C. Linnaeus, 1758' (zoologists always include a publication date, a practice that has recently become more common under the ICN).

How about the species that has to move house to another genus? In the ICN the original author is cited in brackets, followed by the author who placed it in a new genus and the date of the change. The mushroom now known and cited as *Russula caerulea* (Pers.) Fr., 1838 (otherwise known, hideously, as the humpback brittlegill) was named *Agaricus caeruleus** by Christiaan Hendrick Persoon in 1801, but renamed by Fries in 1838 when *Agaricus* was split into many new genera, one of which was *Russula*.

It is even possible to inflict this revisionism on your own species, as with the mushroom *Cortinarius atrocoeruleus* (M. M. Moser) M. M. Moser, 1967. In zoology the latest author is not mentioned at all, but the original author is still bracketed to show that there has been a change in name. However, another indication that more than one author has had a hand in defining a species is the use of *sensu* ('in the sense of'). So '*Cancer pagurus* Linnaeus *sensu* Latreille' refers to Pierre André Latreille's description of *Cancer pagurus* (brown crab), a species originally named and described by Linnaeus.

Quite complicated histories can be conveyed by citations. In '*Cortinarius fulvoincarnatus* Joachim ex Bidaud, Moënne-Locc. & Reumaux 2001', for example, the 'ex' indicates that Joachim described this species (in 1936) but failed in some way to publish it according to the rules, whereupon Bidaud et al. published it again, this time correctly but crediting Joachim with the species.

A thorough author citation includes bibliographical citations as well, which explains hopelessly indigestible offerings such as '*Zanthoxylum cribrosum* Spreng., Syst. Veg. 1: 946. 1825, 'Xanthoxylon'. (*Z. caribaeum* var. *floridanum* (Nutt.) A. Gray in Proc. Amer. Acad. Arts 23: 225. 1888, 'Xanthoxylum')'. There are many more conventions and rules for tracing the authorship of a species, but I think we have seen enough. We should be grateful

---

* The change in ending is purely grammatical.

that author citations are not necessary when writing a gardening column or simple field guide.

Types

During half-time at a meeting on seaweeds at the Natural History Museum in London some years ago, I and my fellow delegates were treated to a tour of the herbarium where thousands of dried algal specimens were stored for study and posterity. Our guide, Professor Juliet Brodie, drew from the shelf a tattered manila folder in which was an ancient, dried specimen of the laver species *Porphyra umbilicalis*. There are probably thousands of specimens of this seaweed stored in dusty cupboards all over the world, but none like this. The one we saw was *the P. umbilicalis* on which all others must depend. It was the 'type specimen' or, more properly, the 'name-bearing type'.

'Types' are of fundamental importance to taxonomy, because they provide the physical objects to which names are permanently attached. A type is an actual specimen – *the* actual specimen, one might say – which fixes the application of a name and sometimes, indirectly, that of a genus or higher taxon. Without a designated type specimen, publication of a new species is (usually) invalid.

Types are the rocks (in palaeontology, literally so) on which taxonomy stands, for they provide a physical basis for what is otherwise merely an idea. Every – well, nearly every – named genus and species will have a dried or bottled specimen stored on some shelf in a university, botanic garden, museum, research institute or, in one instance I know of, a bungalow by the sea. There are, of course, hundreds of thousands of them. Many are contained in jars ('spirit material'); some as microscope slides.

Others reside in cardboard boxes and many (usually plants and fungi) are stuck inside card folders.

The latter, usually called 'sheets', can look like works of art, but they are not museum exhibits despite often being held in museum collections. These are working specimens and, in addition to the copperplate script of the original author, the biro scrawls of later taxonomists can often be seen. Many even have barcodes stuck on them (try this with the early Christian texts found at Nag Hammadi and see what happens to you).

Vast numbers of specimens are kept around the world and only a small proportion are types. Generally they will be indicated as such on the sheet, box or label. Some, like the types of trilobites that I was once shown by Professor Fortey in the Natural History Museum in London, are marked by a little green sticker, while red folders are often used to hold plant types.

Type specimens are continually referred to by taxonomists interested in particular species or genera. Arguments in taxonomy can (sometimes) be settled by reference to the type material because it is a real object and not just a matter of opinion. Unfortunately the modern practice of determining species using DNA analysis derives little support from types, because the DNA of preserved specimens is usually in poor condition.

Types were not explicitly used by Linnaeus, and the idea of employing them on a systematic basis did not arise until the early twentieth century. Now every new species description must specify a type specimen and the place where it can be found (this is usually a museum or similar, but if it is in your attic, provided you say so, that is fine). The type does not even have to be a complete specimen; sometimes it is just a fragment. Palaeontologists have many type specimens locked away in drawers, many of which are, inevitably, mere fragments, though seldom of animals or plants but of rocks.

On rare occasions a drawing or a photograph can take the place of a physical specimen. Shooting one of the last known pair of a newly discovered mammal, for example, would not advance the cause, and certainly not the reputation, of taxonomy, so a photograph would be accepted instead. In zoology, at least, it is the subject, not its drawing or photograph, that is the type. A new species of macaque, *Macaca munzala*, was discovered in north-eastern India in 1997; in the paper that describes it, Sinha et al. denote the type as 'An adult male, photographed by M. D. Madhusudan (Fig 2 Top panel)'.[30]

Tens of thousands of specimens were turned to ash during the Second World War, when the Berlin herbarium was destroyed during a bombing raid (not, as far as I know, on purpose). This has proved to be a long-running disaster for many specialties within the plant kingdom. All of Georg Bitter's African *Solanum* type specimens and eighty thousand sheets, many of which were types, made by the curator of the Berlin Botanical Garden, Ignatz Urban, were lost, among much more. However, in 1929, funded by the Rockefeller Foundation, a Mr J. Francis Macbride had been dispatched to Berlin to take photographs of many of the type specimens for the purpose of making them available to American botanists. He managed to take a remarkable forty thousand images, fifteen thousand of which are now available online.[31] Thus was created the largest collection of iconotypes (picture types), as they are called.*

Even before the twentieth century most naturalists saved a

---

* Lacking official recognition, for the present at least, they must be considered de facto types.

specimen or two of the organism they described and named, and these are often designated as de facto types, although illustrations are often accepted in their place. While this has been very useful, it has been the cause of much embarrassment over the years, when associated specimens are found not to agree with the species described. In 1980, Adrian Pont discovered two specimens of fly labelled respectively *Musca meteorica* and *M. domestica* in the Linnaean collection at Burlington House in London, neither of which was what it purported to be.[32] Similarly, in 1974, Nourish and Oliver found several different species in a single collection of lichens that Linnaeus had named *Lichen rangiferinus*.[33] Fortunately one of them actually was *Lichen rangiferinus*, so the day was saved and it became accepted as the type specimen.[34]

The idea of a type could not be simpler. Unfortunately, the practice of 'typification', as assigning a type is called, is far from simple: human fallibility and the fundamental messiness of nature conspire to make it so. Welcome to arcane alley. First of all, the rules are different in different codes. For example, while types are understandably associated with species, they are also, less understandably, associated with genera. Under the ICN, the type of a genus is the type *specimen* of a single designated species within that genus, usually the first species that was assigned to the genus when it was created. In zoology, the type of a genus is not an actual specimen, it is the type *species* of the genus, not the type specimen of the type species. If the image of angels and pins dances before your eyes at this point, you have my sympathy. But, fine differentiations apart, the fact that genera, too, have types is an important one.

Types come in several flavours and it is necessary (for taxonomists anyway) to know which type of type you are talking about. The most important one, and certainly the easiest to understand, is the holotype. This is the single specimen most closely studied by the naming taxonomist, or the single specimen designated by the author as the name-bearing type. Surprisingly, it does not have to be 'typical'. The fact that it is the one on which the species description is based is all that matters, although obviously a taxonomist is likely to have chosen the most typical specimen if more than one is available.

Where an author has based a description on a collection of specimens (the type series), several of which are cited in the original publication, and no single specimen is labelled as the holotype, any single specimen from this collection is called a syntype. Syntypes are common in earlier collections that take the form of cardboard boxes or envelopes full of samples of the species being named. Nowadays such vague collections are discouraged and a holotype must always be specified by the author. A lectotype is one of the syntypes, which has been subsequently chosen by a later researcher to be the type specimen – a sort of retrospective holotype (the specimens left over after the lectotype has been selected are called the paralectotypes). If an author had specified a specimen as the holotype, those left over would be paratypes.

If you find some zoological paratypes in your box that are of a different sex to the holotype, they are called allotypes. In zoology, male holotypes are generally preferred to female ones, on the sensible principle that there is more to look at without any unseemly poking around inside your specimen. Another commonly used type is the neotype. Things get lost, and some species had no type to begin with, so with much worry and heart searching a brave specialist will designate a new type to define the species name in question. If the existing type material (holotype, lectotype or even neotype) is ambiguous in some way, perhaps missing important characteristics, making it impossible for it to be assigned to a particular species, then an epitype is found and designated. It effectively supplants the previously designated type, but that type must be cited (and kept!). Great efforts are made by biologists to find suitable material for epitypes, some even attempting to collect a specimen from the same location as the original type. Not always successfully, however. The mycologist Kevin Hyde, trying to find a new specimen of the fungus *Colletotrichum circinans*, was disappointed to find that the original site had become a housing estate.

By far the most important types are holotype, lectotype, neotype and epitype, because they unequivocally link a name with a specimen. It is these that are mentioned within the various codes. However, we have barely dipped a toe in the dark and murky waters of types, because taxonomists have come up with terms to cover every conceivable – and inconceivable – eventuality. The distinguished mycologist and taxonomist Professor David Hawksworth lists no fewer than 212 names for different types in his glossary of nomenclature.[35] While some are frankly frivolous, every one expresses a situation that has arisen at one time or another in the attempt to tie a specimen to a name.

Some are special cases of the aforementioned types, such as alloparalectotype, of which I am sure you can work out the meaning. Some have been introduced to service the peculiar requirements of specialists, such as bacteriologists, for whom the original strain of a microorganism kept in culture is called a holocultype. Palaeontologists needed a name for the mirrored impressions left by fossils and settled on countertype. Duplicate parts of the holotype are called isotypes, and the first lucky specimen of a species to have its genitals examined is an aedeotype. A type waiting to be published is a chirotype – presumably because it is 'in hand' – and a type that no one can make head or tail of (sometimes literally) is an ambiguotype.

'Kleptotype' is the name assigned to a type that has been pinched from a museum or institute – a problem that has arisen with lamentable frequency over the years. The story of the museum that was delighted to receive a bequest of a large number of specimens has become legendary in taxonomic circles. The joy of the curators was somewhat short-lived when they realised that the collection consisted entirely of items stolen by their 'benefactor' over several decades. By contrast, a neglectotype is one that has been discovered on a dusty shelf when someone's office was cleaned out years after everyone had thought it lost or stolen. If, in such a case, a neotype has been established during the specimen's sojourn on the shelf, that immediately becomes a neocotype. 'Atypicotype' is the oxymoronic name given to a type that even the author admits doesn't look anything like other examples of the species it typifies, and 'cryptotype' typifies a species described in an obscure journal where no one is likely to find it.

A hoaxotype is simply a fraudulent type, while a heterotype may or may not be fraudulent, being made up of combined bits and pieces from two or more species. The fossils at the heart of the notorious Piltdown Man fiasco were both hoaxotypes and heterotypes. When Charles Dawson discovered the bits of fossilised skull and, eventually, jaw at Piltdown in East Sussex in 1912, he enlisted the assistance of palaeontologist Arthur Smith Woodward from the British Museum. Despite the misgivings of many, Smith Woodward was content that the find represented a 'missing link' between man and apes, noting that the skull was human in shape, if not size, and the jaw indistinguishable from that of a chimpanzee. There was, as it was later revealed, an extremely good reason for these facts. Charles Dawson, the presumed hoaxer, had 'previous' in this area in the form of *Plagiaulax dawsoni*, a 'hybrid' between a mammal and a reptile that he claimed to have discovered in the 1890s.

Here are a few more that I rather like. Artotype: the type of a species that seemed to be new but was in fact an already extant species splattered with paint. Diplomatotype: the type of a species named for political purposes. Taxonomists spend much of their time with squidgy, fragile or otherwise perishable material, hence the ephemerotype, which soon has to be replaced by the neoephemerotype. Copulotype probably requires no explanation beyond the observation that they are nearly always fossils. Finally, I give you the abruptotype, a 'type of a taxon hastily described to meet a project, grant or publication deadline'.[36]

Synonymy and homonymy

When I first took an interest in mycology about forty-five years ago, the meadow waxcap was called *Hygrophorus pratensis*. A few years later it became *Camarophyllus pratensis*, then *Cuphophyllus pratensis*. Its current name (at least the name it had this morning) is *Hygrocybe pratensis*. It is hard enough learning these names once without having to learn the damn things four times. My friend Bryan tends to keep up with these annoyances better than I do, and

I have lost count of the times he has said: 'Well, actually, John, it is *Xerula* now,' or 'That changed to *Parasola* last year.'

It is a constant complaint that Latin names change all the time. Two particularly vexing elements of nomenclature – synonymy and homonymy – partly explain and completely record the phenomenon. Synonyms are different names for the same thing and homonyms are the same names for different things. Synonyms come into being for a variety of reasons and, unfortunately, there are many more synonyms than accepted names cluttering databases and taxonomic papers. The meadow waxcap is lucky in only having half a dozen; some species have many more. The horse mushroom, *Agaricus arvensis*, has twelve synonyms, and the olive, *Olea europeae*, at least seventy, if the varieties are considered to be synonyms. The big one, however, is the common potato. There are only four species of cultivated potato, but they share between them some six hundred names, 596 of which could be considered as being superfluous synonyms.[37]

Synonyms are no-longer-used names given to a particular taxon (species, genus or family). Those in taxonomy are quite unlike synonyms in everyday usage. While you could happily use the words 'domicile' and 'abode' interchangeably, you could not do so with *Agaricus auratus* and *Tricholoma equestre*. The second of these names for what is known as the man-on-horseback[*] mushroom is the only permissible Latin name. However, a permissible name is not considered to be a synonym; only the name or names that it supersedes are synonyms.

Often synonyms occur through a reassessment of a genus, frequently in industrial quantities when a change in the name of a

---

* Another fine example of why I hate imposed common names.

genus automatically generates a large number of changes in species names.* Genera have a tendency to grow as new species belonging to them are discovered. A genus may stay intact as a 'good' genus, but quite often it becomes apparent that the species within it fall into two or more distinct groups, and the genus may be split. The group containing the type species of the genus will retain the original genus name and new genus names will have to be created for the rest. So the old names of 'the rest', which now include an out-of-date generic name in their binomials, will all be synonyms.

This sort of thing can cause weeping and gnashing of teeth among interested parties, as in the notorious case of 'What happened to *Coprinus*' (as it is spoken of darkly by mycologists). Until 2001, the mushroom genus *Coprinus* contained all of the species of fungi known as the ink caps. It was a large and unwieldy genus containing species that really, really did not belong together, and was overripe for the taxonomic cleaver. It was duly divided into four genera: *Coprinus*, *Coprinellus*, *Coprinopsis* and *Parasola*. In any such split the type species of the genus (in this case, *Coprinus comatus*, the shaggy ink cap) must remain in that genus and new names be found for the new genus or genera, unless a proposal is made and accepted to change the type species beforehand. Although very few species were considered to belong to the newly circumscribed *Coprinus*, no such proposal was made and, at a stroke, *Coprinus* was reduced from several hundred species to just a handful. Rather late in the day, a proposal was made to change the type species of *Coprinus* (*C. comatus*) to another species (now *Coprinopsis atramentarius*), which belonged

---

* These are known as 'autonyms'. They are accepted as valid even if they never find themselves in print.

to a larger group. This failed and there are now just a few accepted species of *Coprinus* and a very large number of synonyms that begin with *Coprinus*.[38]

Sadly, most people care little for the niceties of mycological taxonomy, but plants are viewed with considerably more passion. There has been a long-running 'discussion' about the fate of the genus *Acacia*, which, like *Coprinus*, was split into several genera. One article on the subject was admirably entitled 'Wattle happen to Acacia?', the golden wattle being a member of the genus (*Acacia pycnantha*). One of the new genera was destined to retain the name *Acacia*, a name that had been attached to African species since Dioscorides in the first century. The rules dictated that it should be the group that contained the type species, *Acacia nilotica*, which was from Africa.[39]

However, more than a thousand of the fourteen hundred species within the existing circumscription of *Acacia* are found in Australia, and it was suggested that these, or at least those among them that fitted the new taxonomy, should inherit the name. Feelings ran high, in part because the golden wattle is the botanical emblem of Australia and the genus as a whole much loved in that country. The matter was finally decided in 2011 at the 18th International Botanical Congress, held, pointedly, in Melbourne, when a special provision was passed for the Australian species to retain *Acacia* despite some serious and heartfelt opposition from some delegates.[40] 'Australia Wins Acacia Wars' was one news headline celebrating this break with convention. It was, and still is, considered a gross miscarriage of natural justice by many European botanists, many of whom are still in mourning.

The above examples are entirely about what to call things, not where they belong. However, a more worrying circumstance comes about through differing taxonomic views. Names are tied to a taxonomic viewpoint (a circumscription), and if a species (based on a single type specimen or series) is placed in one genus by Dr Smith and in another by Dr Jones, it will inevitably have two different names. Both taxonomists will consider the species to have one good name and at least one synonym, but will hold reverse opinions on which is which. The tropical tree *Adenanthera falcataria*, for example, has three other names, all nomenclaturally 'correct' – *Albizia falcataria*, *Paraserianthes falcataria* and *Falcataria moluccana* – each one reflecting the points of view of particular specialists. Differing views on taxonomy that result in several names for the same thing, although scientifically understandable, are a great irritation to anyone who needs to use those names. The average writer of reports on conservation, for example, would be implicitly accepting one taxonomic view over another every time he or she used a disputed name; something they could not be expected to do. All that can be hoped is that the taxonomists concerned sort it out between themselves or, failing that, that one or more of them dies.

When a new name is given to an already named type specimen, as in the above examples, the resulting synonym is described as a homotypic synonym.* However, synonyms can also come about by the independent naming of a species more than once, based on more than one type specimen. This is termed heterotypic

---

* An example for clarity: the old name *Coprinus disseminatus* is a homotypic synonym for *Coprinellus disseminatus* (the new name), but it is still based on the same type specimen.

synonymy. Typically, someone discovers, names and describes a specimen in ignorance of the fact that it has been discovered, named and described already. The crucial difference about this form of synonymy is that it arises from the establishment of a new type for an already named species. One of the challenges when naming a species is to avoid this trap; no small feat, because a name and associated description may be hidden away in some ancient journal or book. With online data now available, it is a simpler task than it once was. However, as a glance at the collected literature on any one grouping of organisms would confirm, relying upon online research alone is perilous. The library devoted to fossil species at the Natural History Museum in London, for example, is the size of the average small town library and most of what it contains is not available in electronic form. Experience, research and consultation with colleagues are needed in order to avoid this embarrassing and time-wasting error.

A subtly different problem occurs when a new species is named in the full knowledge of the existing literature and type material by a taxonomist who considers the discovery to be sufficiently distinct to warrant a new name. This commonly occurs in the recognition of subspecies, varieties and forms, which are often described in a fit of pickiness by taxonomists and then dispensed with later by others as annoying synonyms. Again, there are two type specimens for one species. This happens with tiresome regularity: five of the eleven synonyms for the horse mushroom, *Agaricus arvensis*, are varieties or forms.

Synonymy is a contentious issue because people get very fond of names, especially those that they have coined themselves. One person's synonym may be another's preferred name. A mycologist friend of mine, complaining about another mycologist's opinion on the nomenclatural status of species dear to his heart, exclaimed to me that 'he knows nothing about the Cortinariaceae, I tell you. Nothing!' The renowned mycologist Peter Orton was an authority on a particularly tricky (and dull) genus of fungi, *Psathyrella*. The only other person I know of who could find a place in his heart for *Psathyrella* is a Dutch mycologist, Emile Kits van Waveren. In 1986, Dr Kits van Waveren published a monograph on *Psathyrella*, which Orton reviewed in the *Transactions of the British Mycological Society*. Orton began his review with a paragraph damning it with faint congratulations and praise, then systematically trashed it over the remaining several pages. Orton was a notorious splitter (making three species out of one), and reserved the bulk of his ire for Kits van Waveren's lumping tendencies (making one species out of three). The names discarded by Kits van Waveren as synonyms were considered by Orton to be good names that had been ill used (I make no judgement either way on who might be right).* Lumpers and splitters effectively conjoin to breed synonyms. The splitter divides one species into two, both of which may be accepted for a while as good species. The lumper then recombines the two species, and a bouncing new synonym is delivered into the world.

There is nothing wrong with this sort of thing. It is just science, and if two or more names exist for an organism, either because of disagreements at species level or different views of taxonomic placement, then we just have to wait to see which prevails. One might imagine that the international bodies that write and manage nomenclatural codes would adjudicate in such

---

* With molecular techniques finding ever finer differentiations, the splitters are in the ascendant.

matters, but their primary task is to decide on technical, rule-based issues, only occasionally and in special circumstances involving themselves in scientific judgements.

Homonyms

Homonymy, another taxonomic annoyance, is the application of the same name to two different species or genera. Again this happens less frequently in these days of electronic searches (even if a description is hard to find, the name is unlikely to be), but in the past it was quite an undertaking to ensure that the name you wanted to use had not been used before. Homonyms formed in this way are referred to as primary homonyms. *Oenanthe* is a genus of plants and a genus of birds, but they are not considered to be homonyms because they are names assigned within different codes (although using a generic name already accepted by another code is now discouraged). More striking still because it is an entire name is *Prunella vulgaris*, which once managed to be both a plant (self-heal) and a bird (the dunnock), but again these are not considered to be homonyms.

A fairly simple example of homonymy that is not acceptable comes from the garden. Charles Jeffrey, in his *Introduction to Plant Taxonomy*, cites the case of *Viburnum fragrans* Bunge. In 1831, Bunge named his specimen of viburnum *V. fragrans*, unaware that *V. fragrans*, as a name, had already been used for another species described by Loisel in 1824. Bunge's species was therefore a homonym and illegitimate. It was subsequently redescribed by William Stearn in 1969 and is now known as *V. farreri*.

In addition to homonyms caused by taxonomists not paying close attention to the literature, there are those created when, in a frenzy of lumping, taxonomists combine two or more genera to form one. Since all of the original species will now share a generic name, any that also share a specific epithet will produce a homonym. Two species of fly, *Frontina acroglossoides* and *Eophrissopolia acroglossoides*, faced this problem when both were placed in the new genus *Chaetogaedia* and one of them had to be renamed. This phenomenon is known as secondary homonymy.

Principle of priority

If you have two names for something, or two somethings for one name, which one wins? The main aim of taxonomy is to provide stable names for all life on Earth, and the 'principle of priority' is an essential tool in achieving this. The principle states that the name of any organism should be the earliest name that was published in accordance with the rules prevailing at the time, and it removes from troublesome circulation the accumulated super-fluous synonyms and homonyms described above. Unfortunately, this simple rule determining which names live and which names die can occasionally cause a great deal of collateral damage, changing long-established names when an older name is discovered. It is simple in theory but hopelessly complicated in practice, and a moment's thought will raise questions such as 'What is meant by earliest?' and 'What names: genus, specific epithet, binomial, higher taxa?'

The answer to the first question at least is easy, if overlong. For zoology, it is the earliest date, starting from 1 January 1758, the year in which the tenth edition of Linnaeus's *Systema Naturae* (see p. 191) was published. For botany, it is 1 May 1753, the publication date of the first volume of his *Species Plantarum*. Bacteriologists once used the 1753 date, too, but in 1980 they effectively ditched all their names and started again from 1 January 1980.

Biological nomenclature is more exception than rule, so it is no surprise to learn that there are more starting dates than these three. The exceptions are all within the botanical code (ICN). For example, all mosses (except, for some reason, the sphagnums) start their nomenclatural life in 1801. The starting dates for fossils are all over the place, although algae are the worst offenders, with five dates to choose from. Names published prior to *Species Plantarum* or *Systema Naturae* are termed 'pre-Linnaean' and therefore invalid. Even some of Linnaeus's own names are 'pre-Linnaean'; if he published them only before 1753 or 1758, they cannot be used.

Among the thousands of little brown mushrooms that lurk in woodlands for the sole purpose of taxing the patience of mycologists are those of the genus *Inocybe*, a name first conceived in 1863 by the great Swedish mycologist Elias Fries. Prior to this, in 1821, he described a species that he called *Agaricus calamistratus*. In 1829, another mycologist, Wilhelm Gottfried Lasch, named and described a species that he called *Agaricus hirsutus*. In 1872, Lasch's species was moved from *Agaricus* to *Inocybe*, resulting in *Inocybe hirsuta*. Both 'species' turned out to be the same, amounting to three potential synonyms. *Inocybe* was now an established genus so the species stayed in *Inocybe*, but the specific name *hirsuta* was a young pretender that was duly deposed in favour of the earlier, rightful *calamistratus*. The species was renamed *Inocybe calamistrata* in 1878 by Claude Casimir Gillet, a name it has retained ever since. This is a reasonably neat example of the principle of priority in action.

Publication dates are viewed in strict chronology, so a name published in 2010 would trump one published in 2012. A new species (and genus) of amoeba, *Hyalolampe fenestrata*, published in 1869 by Richard Greeff, was quickly consigned to inglorious synonymy when it was discovered that the same species had been published a few days earlier under the name of *Pompholyxophrys punicea* by William Archer. I prefer Greeff's name, which means something like 'lamp with clear windows', to Archer's 'purple blister with an eyebrow', but opinion is of no importance and the blister wins. Priority takes, er, priority, even if the original name was absurd and crying out to be replaced with a much more suitable name. This was an irresistible temptation to many taxonomists prior to the establishment of this rule and thus ultimately instrumental in its inception.

Biological nomenclature possesses a richness of embarrassments in unsuitable names, but because of the principle of priority they cannot be changed. Taxonomists often receive through the post dried and dilapidated specimens, which they are required to describe and name. They are remarkably good at this, considering what they have to work with. However even the most thorough can be confounded in their efforts. *Lichen aromaticus* (now *Toninia aromatica*) has little in its name to describe it, apart from telling us that it is a lichen and that it has a nice smell. Well, it *is* a lichen, but it has no smell at all; the misnomer was a result of Messrs Turner and Borrer sending the specimen to Sir James Edward Smith in a perfumed envelope.

The principle of priority applies up to the level of family under the ICN and just to genus under the ICZN. But it is the binomial itself, the individual species name, which falls most frequently to its judgement. This is chiefly because there are, of course, many more of them than there are genera or families. However, one example of a genus fighting off a usurper is that of the hammerhead sharks. The impostor started life as the specific epithet in

Linnaeus's name for the smooth hammerhead shark, *Squalus zygaena* ('shark shark' in Latin then Greek). As with many old epithets, the genus in which it found itself was split and the epithet promoted to that of one of the new genera: *Zygaena*. Unfortunately this was already a genus of moth, established in 1775 by Johann Christian Fabricius, and *Zygaena* the fish had to go. Well, almost – it went back to its old job as the specific epithet of the hammerhead shark in *Sphyrna zygaena* (hammer shark).

Priority has had the unfortunate side effect of removing from circulation well established and familiar names. For example, the binomial *Viburnum fragrans* enjoyed uninterrupted popularity until Stearn found out that it was a homonym in 1969. Unfortunately by that time it had become a much-loved garden plant and Stearn's renaming of it, while quite proper by the rules, was not at all welcomed. There is a natural resistance to the revision of long-accepted names, especially such euphonious ones as *Viburnum fragrans*, but good names fight back – plant catalogues often sell this species as '*Viburnum fragrans (V. farreri)*'. One name that refuses to lie down and die is *Brontosaurus*. It was synonymised with *Apatosaurus* more than a hundred years ago, not long after it was first devised, and its steadfast refusal to leave was duly vindicated in 2015 when it was accepted as a separate genus.

Another serious problem caused by the principle of priority is the resurrection of lost names found by overenthusiastic researchers on dusty shelves in museum libraries. These are, generally speaking, not lost treasures but lost liabilities and their rediscovery causes many headaches. Occasionally a large cache of them is found in a single publication and chaos threatens. The highly colourful French-German-American naturalist Constantine Samuel Rafinesque,* for example, provided binomials for 6,700 species of plants and a large number of animals in the early nineteenth century. Few were accepted in his day because his approach to taxonomy was considered by his peers to be slapdash and didactic. But many of his species, not accepted at the time, were good and take priority.

In recent years, Rafinesque's works have provided much soul-searching, head-scratching and bad-tempered argumentation among taxonomists. The slightly masochistic quest for lost Rafinesque binomials has become what Leonard Warren, in his book on this disregarded savant,[41] describes as 'a minor cottage industry'. Some of his names have been reinstated and some have always been accepted, but they form a tiny part of his output. Rafinesque is unusual in having produced such a vast collection of troublesome names, but individual unwelcome binomials reappear all the time.

In fact, there is now a mechanism to protect names that are so well accepted that to change them would be more trouble than it is worth. This is what happened to the Australian species of *Acacia* and not to the mushroom genus *Coprinus*. Such a name can be declared a *nomen conservandum* (botany) or 'conserved name' (zoology). It is formally accepted that even though it is a junior (later) synonym or a homonym that fails the priority test, it can be retained. This formalised triumph of common sense over rule-book diktat has saved many species and genera over the years. For example, the common and tasty fairy ring champignon, *Marasmius oreades*, was threatened, together with all other *Marasmius* species, with the name *Microphale*, which was discovered to have priority.[42]

---

* Rafinesque is commemorated in the daisy-like flower genus *Rafinesquia*.

To the great relief of all those facing the loss of a familiar and cherished name, *Marasmius* was conserved.

While conserving generic (and family) names was long seen as an acceptable concession to common usage, conserving specific epithets was not. However pressure grew from agricultural and horticultural botanists who were understandably tired of constantly changing names. In 1981, it was agreed that the specific epithets of economically important plants could be conserved. The cause célèbre of the interested parties was wheat, *Triticum aestivum*, which, under the principle of priority, should really have become *T. hybernum*. But everyone was used to it and wanted to keep it,[43] and in 1988 it became the first specific epithet to be conserved under the botanic code.

The tomato, which has a chequered taxonomic history, has the conserved name *Lycopersicon esculentum*. Because of priority it should really be *Lycopersicon lycopersicum*, but botanists do not allow tautonyms (two identical words in one binomial) and don't much like binomials that look as though they want to be tautonyms, so *Lycopersicon esculentum* it remained. Incidentally Linnaeus called it *Solanum esculentum* and some authorities prefer this name, considering that the tomato should not have been moved to the new genus *Lycopersicon* in the first place.[44] [45] Such is the convoluted world of nomenclature. I often tell people not to worry about constant changes in names because if you live long enough they revert to what they were when you first learned them. Evidently there is some truth in this bit of wishful thinking.*

---

* The various bodies that administer the codes, evidently despairing of the endless problems caused by a system based on rules, are seeking to move to approved lists agreed by the taxonomists themselves. Now, for example, the Approved Lists of Bacterial Names and the List of Protected Generic Names for Fungi have come into existence.

The fate of names discovered on dusty shelves and then rejected in favour of conserved names casts a light on taxonomists' fondness for arcane terminology. When a species is first described, it is a *species nova*. The name then passes some time as a *nomen approbatum* (approved name) then, perhaps, its light failing to shine, a *nomen neglectum*. Over time it sinks utterly into the shadows and becomes a *nomen oblitum* (forgotten name). It is rediscovered and lives in hope of becoming a *nomen reclitum* (revived name), only to spend months or years awaiting the judgement of an international committee as a *nomen rejiciendum propositum* (proposed rejected name). If things go badly, it ends its days a *nomen rejiciendum* (rejected name) and disappears forever into obscurity, bitter and disappointed.

Nomina rejicienda do, however, continue in ghostly form, because the name can never be used again. The name it hoped to replace is now called a *nomen conservandum*. *Nomina* of various flavours infest nomenclature. *Nomina dubia* are quite common, appearing every time the type material is suspect; *Aachenosaurus multidens* (see p. 47) is a prime example of a *nomen dubium*. My favourite *nomen* is *nomen nudum*, a name that has been mooted in a publication but without any accompanying description. Gavin Prideaux from Flinders University in Adelaide kindly told me about a species of an extinct tree kangaroo he is about to name. Since his publication date is likely to be after mine, my mention of the species name in this book predates his and will establish, for a while, a *nomen nudum*. Here we go: it is called *Bohra dagon*.

Tautonyms

Names such as *Crangon crangon* (brown shrimp), *Bufo bufo* (common toad), *Naja naja* (Indian cobra) and, deliciously, *Indicator indicator* (greater honeyguide) are fairly frequent within zoology and completely absent from botany, mycology and other disciplines covered by the ICN. These tautonyms (repeating the same word for genus and specific epithet) would occur in botany, but are simply not now allowed in the rules, so it is only zoologists that have access to these poetic delights.

As far as I can tell, tautonyms are never coined afresh, so how do they arise? It is the nature of genera to be reviewed in the light of new understanding of relationships, notably those that have accumulated a large number of species over the years. As a result, they are often split into two or more genera, for which new names must be found. Where better to look for a new name than among the specific epithets within a new genus? In this way, many specific epithets undergo 'promotion' to generic status – this has happened hundreds, perhaps thousands of times. For example, the gorilla was once called *Troglodytes gorilla*, but in the mid-nineteenth century Isidore Geoffroy Saint-Hilaire became convinced that it should belong in a new genus. Tautologies were little used in those days and he gave it the new name of *Gorilla savagei*. Priority being a firm rule in taxonomy, the new specific epithet was later replaced with the old one and the species became *Gorilla gorilla*.

The story does not end here, however. When a species acquires a subspecies, the binomens (as zoologists call binomials) become trinomens to distinguish between them. There are, the taxonomists tell us, more than one type of gorilla, and the Western gorilla has now been divided into two subspecies: one is *Gorilla gorilla diehli*, and the other, to the glory of the world, is *Gorilla gorilla gorilla*, the original species always displaying its subspecific status by the repetition of its specific epithet. In similar fashion, we have earned *Chloris chloris chloris* (greenfinch), now sadly a synonym of *Carduelis chloris*, and the previously mentioned *Troglodytes troglodytes troglodytes* (wren), a bird that is shorter than its name.

Ornithologists are the most enthusiastic purveyors of tautologies – a quick search has unearthed more than a hundred birds so adorned, as well as about fifty among the mammals, although there will be many more. They are some of the most attractive of all Latin names: *Oenanthe oenanthe* (wheatear), *Alle alle* (little auk), *Buteo buteo* (common buzzard), *Boops boops* (bogue, a species of seabream) and *Natrix natrix* (grass snake). I particularly like the eastern kingbird's impressive name: *Tyrannus tyrannus* ('absolute ruler', times two). There is temptation to think that *Crex crex* (corncrake) and *Pica pica* (magpie) are repeated just for onomatopoeic effect, but it is not so.

As noted, botany and allied disciplines avoid all this by simply banning tautonyms. When Linnaeus's genus *Gentiana* was divided into many new genera, one of them, *Gentiana centaurium*, had its name co-opted for the new genus *Centaurium*. However, it could not become *Centaurium centaurium* because the rules do not allow this. Instead the plant was given the old epithet of Gaspard Bauhin (see p. 173), *minus*, to become *Centaurium minus*. The botanical ban applies only if the two words have exactly the same spelling, so the shrub *Ziziphus zizyphus* is 'technically' acceptable. Sadly taxonomists are not always the fun-loving creatures we have come to know, and the former name, despite being legal under the rules, found itself banished in 2006 for being too much of a near-tautonym,* to be replaced by *Zizyphus jujuba*. Fortunately one of the most appealing of all names has not suffered so dismal a fate: the salak palm, *Salacca zalacca*, has remained unmolested.

---

* Known as a virtual tautonym in the trade, this term also covers the situation where two words have the same meaning, as in *Bos taurus* (see p. 22).

Equisetum palustre brevioribus
foliis polyspermon

CHAPTER VI

The HISTORY of NAMING

*And out of the ground the LORD God formed every beast
of the field, and every fowl of the air; and brought them
unto Adam to see what he would call them: and whatsoever
Adam called every living creature, that was the name
thereof. And Adam gave names to all cattle, and to the fowl
of the air, and to every beast of the field.* Genesis 2:19, 20

DESPITE THE JEALOUS CLAIM OF ONE NOTORIOUS OCCUPATION, IT can be forcefully argued that taxonomy is the oldest profession. The irascible God of the Old Testament took a poor view of humankind, His commands largely being those of proscription rather than prescription. But one thing he did ask us to do, in fact the very first thing he ever spoke to us about in the Garden of Eden, was to name the animals.

God is merciful, so it is said, and it is perhaps for this reason that, while asking Adam to provide the animals with names, he did not require him to put them in any order. Yet the list he gave to Adam – cattle, fowl and beasts of the field – hints at a kind of order, a folk taxonomy that would work for all time at a practical level and that would serve as the only taxonomic paradigm for

generations. It was rather restricted, but the one thing its creatures
have in common is that they are edible. Pragmatism was (and,
with folk taxonomies, still is) everything when it came to classification and, with a handful of classical exceptions, it dominated
until the Renaissance.

Finding the true order (see p. 211) of the living world was of
little practical use until modern times. Fennel (*Foeniculum vulgare*)
is an excellent accompaniment to fish and calms the stomach;
hemlock water-dropwort (*Oenanthe crocata*) tastes awful and will
have you dead in three hours. That is enough information for
most people; the fact that both are members of the family
Apiaceae and thus closely related is of no particular concern.
Classifying organisms according to utility is no more scientific
than putting them in alphabetical order, but both are at least some
sort of order. The search for the true order would fall to Adam's
descendants many millennia later.

In their endless struggle to avoid the tragic end of the Dodo
and countless millions of other ill-fated species, all animals are
equipped with a sense of the order of the world in which they find
themselves. For anything above the size of an amoeba (and maybe
even for them), fellow organisms may be categorised simply
enough: the good, the bad and the indifferent. That is, things that
you can eat, decorate your nest with, shelter under, use to cure a
headache; things that will eat you, poison you or bring you out in
a nasty rash; and, so much less interesting, everything else.

One would hope that such considerations were largely
beneath civilised modern man, who should appreciate organisms
for their beauty or scientific interest, but it is not really so. On the
many fungus forays I run every year, I will name a fungus we have
found, provide details of its relationship to other species, its

structure and its mode of life, and express my delight in its beauty. 'Yes, John, but can you eat it?' is, after a polite pause, the usual response. If it is edible, it becomes a thing of great interest; if poisonous, a thing of fascination. If neither, we move on to the next fungus.

Early plant lists

As soon as some sort of writing developed, people began to keep lists of plants, and in those that survive from early times this attitude of pragmatism prevails. By far the commonest class of species listed was those that had medical interest. Libraries seldom benefit from a fire, but in an act of constructive vandalism in 612BC, the Babylonian invaders and destroyers of the Assyrian city of Nineveh inadvertently baked 22,000 clay tablets that had been held in the library of King Assurbanipal. Preserved for a delighted posterity, they were exhumed in the mid-nineteenth century by the adventurer Sir Austen Henry Layard, and the cuneiform writing of 660 of these tablets and others found elsewhere painstakingly examined and translated by Reginald Campbell Thompson. They consisted in part of herbals, listing some two hundred species of plants, 150 of which Campbell Thompson was willing to have a shot at assigning a modern name to. From aloes to elder, fennel to rosemary, saffron to wormwood, the list includes plants that are familiar still, although squirting cucumber is a little hard to come by these days.[46] Most astonishingly, while the list was discovered two and a half millennia after it was baked and buried, Campbell Thompson revealed that many of the tablets were copies of Sumerian texts written two and a half thousand years earlier, making them close to five thousand years old.[47]

Just as long ago (3000BC) in China, there lived, or at least was purported to have lived, the second of the mythical emperors, Shen Nung, the 'divine gardener'. His mostly mythical status is forcefully suggested by the story that he had the body of a man and the head of a bull, but he is nevertheless credited with authorship of *The Divine Husbandman's Materia Medica*, a work that catalogues 365 species of medicinal plants and that was moved from oral tradition to written tradition in the first century BC. Legend tells us that he experimented with the medical properties of plants by the simple and ultimately doomed expedient of eating them himself. He managed to recover from the first sixty-nine poisonings, the seventieth evidently being his last.[48] [49] [50] He is also credited with the discovery of tea and the invention of acupuncture, for which we can be respectively grateful and disappointed.*

One of the earliest, and certainly one of the most beautiful memorials to the ancient world's interest in medical plants is the Ebers Papyrus. For once we have a reasonably reliable date, because someone thoughtfully provided it on the back, placing it in the ninth year of the reign of the Amenhotep I, the second Pharaoh of the eighteenth dynasty of Egypt (circa 1534BC). It is a magnificent scroll twenty metres long, consisting of 110 pages of hieratic script (a stylised relative of hieroglyphs). It came to light at Thebes in the mid-nineteenth century, although precisely where it was discovered is rather obscure. However, it is believed to have been found wedged, rather uncomfortably, between the legs of a mummy, the deceased displaying an impiously sceptical

---

* Although a consummate cynic, I was prepared to try acupuncture when suffering from a bad back. I was as honest as possible about the effects, which were, it transpired, zero. The lady who did it for (to) me was terribly disappointed, saying that I was her first total failure.

view of the afterlife in bringing a medical textbook with him. The scroll is as much magic as medicine, but it shows considerable knowledge of the human body and mentions in its seven hundred-odd drug recipes many plants known to have true pharmacological value. Among these are parsley, poppy, garlic, aloe vera* and deadly nightshade. Not all of the three hundred ingredients included are plants: such things as sulphur, iron oxide, honey and crocodile dung also receive a mention.[51][52]

There are many more extant ancient writings that provide lists of plants and indeed animals, and undoubtedly many more that have been lost or remain undiscovered. But they are all utilitarian in nature, showing little interest in the arrangement of organisms. The legacy for taxonomy of these herbals and those who wrote them is in their exploration of nature, for one cannot classify what one does not know. The old herbalists must, nevertheless, have noticed similarities between otherwise clearly defined species, as is suggested within the Ebers Papyrus, where the writer speaks of the 'celery of the plain' and the 'celery of the hills'.

Aristotle (384–322BC)

It was during the classical period that the natural world first became an object of study rather than exploitation, and many modern biological concepts find their first flowering at this time. Here were the first scientific descriptions of plants and animals and the first faltering steps towards a system of classification.

It has been said that Aristotle was the last man to know everything there was to know in his time. The list of subjects he

---

* One of very few species whose common name is the same as its Latin name. Another is *Boa constrictor*.

studied or wrote on reads like a university prospectus: aesthetics, anatomy, astronomy, embryology, ethics, geography, geology, government, literature, metaphysics, meteorology, physics, politics, psychology, rhetoric, theology and zoology.

All biology starts with Aristotle. Taxonomy is no exception. Among the thirty or so of his works that have survived the trials of time, three or four deal, at least in part, with taxonomy.

Aristotle was a Macedonian, born in Stagira, located in the three-fingered peninsula of Chalcidice. His father, Nicomachus, was physician to Amyntus III, so Aristotle was neither particularly lowborn nor highborn. But as a student of Plato and a tutor and friend of kings (including Alexander the Great), by the end of his life he could out-namedrop anyone. Aristotle studied with Plato in Athens for twenty years, and when his mentor died, he moved to Assus in Anatolia (in present-day Turkey). Here he studied the marine environment, but he ran into political trouble by assisting Hermias, the ruler of the city, to negotiate an alliance with Macedonia, angering the Persian King, Artaxerxes III. The King, not known for his forgiving nature, had Hermias executed, and Aristotle fled to Lesbos.

It was on Lesbos that much of the research for his three works of natural history was done. Not that he did it alone – he had a considerable number of assistants, one of whom was Theophrastus (see p. 154). Aristotle stayed on Lesbos for a year or two, then, after a thirteen-year absence, returned to Athens, where he established his own school at the Lyceum. It was called the Peripatetic school, the name meaning 'walking around', owing to Aristotle's habit of teaching on the hoof. *

---

* This was an early pre-echo of the similar habit of Linnaeus.

Although, as we have every right to expect, Aristotle was given to much philosophising, he was considerably more practical in his approach than his master. Plato thought that knowledge could be acquired by just thinking really, really hard; Aristotle believed that knowledge is gained through enquiry. His major work on biology is known as *Historia Animalium* (*The History of Animals*). Alongside this book are two treatises, often described as short (they aren't, I have read them): *De Generatione Animalium* (*The Generation of Animals*), and *De Partibus Animalium* (*The Parts of Animals*). These, together with bits of other writings, provide us with Aristotle's view on the organisation of the natural world.*

The History of Animals is a first-class read and quite extraordinary for its time. There had never been anything like it before. Aristotle presented the natural world to people in a new way, describing the lives and loves of five hundred species. He was, in short, the David Attenborough of his time. The book ranges from the ridiculous – 'the elephant, which is reputed to enjoy immunity from all other illnesses, is occasionally subject to flatulency' – to the sublime: 'The dolphin and the whale, and all such as are furnished with a blow-hole, sleep with the blow-hole over the surface of the water, and breathe through the blow-hole while they keep up a quiet flapping of their fins; indeed, some mariners assure us that they have actually heard the dolphin snoring'. Most of his many, many fascinating observations on the animal world are accurate and perceptive, but he does fall down seriously on occasion, for example in believing that fish are sometimes generated from mud.†

---

* Aristotle is also believed to have written on plants, but these works appear to be lost.

† Presumably a reference to lungfish.

Aristotle's taxonomy

Academics have expended much ink disagreeing with one another about Aristotle's taxonomic legacy: what and how much his contribution was, and even if he left us anything of lasting value at all. What is clear is that Aristotle did not formulate anything like a complete taxonomic system. But he laid some substantial footings, a few hefty foundation stones, several bricks from higher up the building and, to extend a poor metaphor further, a few decorative architectural flourishes (such as the word *entoma* for insect, from which we get 'entomology').

I suspect that Aristotle, the inventor of Logic, would have liked nature to be much neater than it is. But in studying it at first hand he found a fundamental truth: biology is endlessly, endlessly messy. Unlike physics, mathematics and line dancing, it has no principles, save the one of which he was not aware – common descent (see p. 199). Aristotle thus quickly dispensed with any idea of the dichotomous division beloved of Plato. Any scheme in which each category is subsequently divided into two falls flat on its face in quick order. Someone could start with the nice neat distinction between animal or plant; then, choosing animal, divide into flying or terrestrial; then, choosing flying creatures, turn to wings, which could be feathered or membranous; and so on, ending up with, say, a bee. Unfortunately this simple series has already left out, at stage two, anything that lives in water and, at stage three, bats. It is possible to get around this by using what Aristotle called 'privation' – for example, winged/not winged, feathered/not with feathers – but this leads to unwieldy, lopsided hierarchies and an even worse mess than the first method.

[148]

Thus Aristotle never attempted to produce a consistent taxonomy; he was faced with far too many anomalies. But he did, almost inadvertently, sketch out what one should look like. To some extent, what has been interpreted as taxonomy in his works is just a method of avoiding repetition. As he indicated, if one first establishes a group of animals that are live-bearing quadrupeds, one does not have to say for every animal in that group that it gives birth to live young and has four feet. The result of this was to produce several overlapping taxonomies. So, in addition to birds being included in the category of 'animals with blood', they were also found within those 'creatures that fly'.

In as much as he attempted an overarching structure at all, Aristotle decided to go with common sense. In *The Parts of Animals* he wrote: 'the proper course is to endeavour to take the animals according to their groups, following the lead of the bulk of mankind'. His main division was between those 'with blood' and those 'without blood': the vertebrates and the invertebrates (although he did not refer to them as such). The invertebrates he divided into 'Molluscs' (sea squirt, octopus), 'Malacostraca' (crayfish, crabs), 'Testacea' (oysters, snails), 'Arthropods' (insects, spiders) and also creatures such as coral and sea fans – although he thought they might be a halfway house between animals and plants. It is hardly surprising that he was unaware that oysters and snails are molluscs, too, or that sea squirts were much more closely related to humans than squids.

Aristotle's vertebrates came in five flavours too: live-bearing quadrupeds (mammals), egg-bearing quadrupeds (reptiles and amphibians), birds, fishes and then whales (cetaceans). He did not seem to realise that whales were mammals, but hats off to him for not thinking they were fish. Of course, Aristotle could not

know of the extraordinary exception to the first of these classes –
the egg-laying mammal, the duck-billed platypus – but he
certainly knew about the viper, a live-bearing, legless reptile. In
this alone he would have realised and accepted the limitations of
his simple scheme. His hierarchies seldom stretched very far. He
stated, for example, that live-bearing quadrupeds should be taken
'one by one . . . lion, stag, horse, dog, and so on'. He was prepared,
however, to grant 'horse' what we would now consider 'family'
status, dividing it into horse, ass and hemioni (wild Asian ass –
although literally it means half an ass).

As well as creating the idea of a biological taxonomy,
Aristotle also provided four of taxonomy's most important
concepts: genus (*genos*), species (*eidos*), essence (*essentia*) and
differentia (*diaphora*). His uses, however, of the first two words
are as general terms, not the specialised ones of today. A genus
for him was any class (in the ordinary sense of the word) and a
species was any member of that class. So what for us is the order
Carnivora, within which, among others, we find the family
Felidae, would to Aristotle be the genus Carnivora containing
the species Felidae. Of course, in the case of the species *Felis
catus* within the genus *Felis*, the modern usage matches that of
Aristotle.

The differentia (both for Aristotle and present-day taxonomists) is what distinguishes the members of any class from one
another. So, to take a modern example, the ciscaucasian hamster
and the golden hamster are both small, furry creatures with short
tails, which belong to the genus *Mesocricetus*, but they are differentiated by the former having cute dark bands around its chest
and living peacefully in Russia while the latter spends its time
running around a treadmill and ungratefully biting the finger of

its owner. I am sure there is a much more scientific differentiation, but I think this makes the point.

Linked intimately to the concept of 'species' is the slightly more slippery idea of 'essence'. This is essentially, well, essentialism – one of Plato's most famous creations, inherited by Aristotle in a mild form. The essential nature of any entity, be it a class or an individual, is the attribute or attributes it must possess to be what it is. So, to be a rabbit you need long ears, a fluffy tail, an insatiable carnal appetite and all the other characteristics that make up a rabbit. If you lack any of these (save by accident – ears blasted off by Mr McGregor, for example), you would not be a rabbit. To be the carrot family you need to be the class of all plants that, among other things, produce five-petalled flowers in umbels (sprays). While genus, species and differentia are important ideas in modern taxonomy, essence, much employed by naturalists from the Renaissance until the early nineteenth century, has fallen seriously out of favour.

The scala naturae

After providing his new science of taxonomy with four important concepts, Aristotle proceeded to shoot it in the knee, if not the head, with a fifth: the *scala naturae* (see p. 14). In *The History of Animals* he wrote:

> *Nature proceeds little by little from things lifeless to*
> *animal life in such a way that it is impossible to determine*
> *the exact line of demarcation, nor on which side thereof*
> *an intermediate form should lie. Thus, next after lifeless*
> *things in the upward scale comes the plant, and of plants*
> *one will differ from another as to its amount of apparent*
> *vitality; and, in a word, the whole genus of plants,*

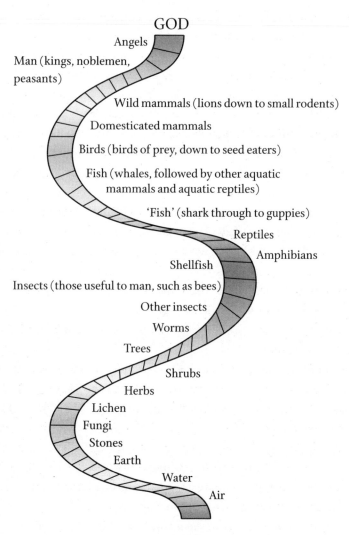

GOD

Angels

Man (kings, noblemen, peasants)

Wild mammals (lions down to small rodents)

Domesticated mammals

Birds (birds of prey, down to seed eaters)

Fish (whales, followed by other aquatic mammals and aquatic reptiles)

'Fish' (shark through to guppies)

Reptiles

Amphibians

Shellfish

Insects (those useful to man, such as bees)

Other insects

Worms

Trees

Shrubs

Herbs

Lichen

Fungi

Stones

Earth

Water

Air

A rough idea of the *scala naturae* drawn from several sources. Ideas on precisely how creation was arranged in ascending order of complexity and importance varied from author to author. Each group shown here will have its own, often intricate, ranking.

> *whilst it is devoid of life as compared with an animal,*
> *is endowed with life as compared with other corporeal*
> *entities. Indeed, as we just remarked, there is observed in*
> *plants a continuous scale of ascent towards the animal.*[53]

These might appear to be unexceptional observations, but a considerable amount of elaboration in later centuries, and the human tendency to transform observation into principle by squeezing subsequent facts until they fit, set taxonomy on a path from which only Darwin could save it (see p. 232). Nevertheless, it is a beguiling principle and one that even today has a powerful sway. We still talk of 'higher' and 'lower' animals; most people consider plants to be less developed than animals – rather than what they are, which is just different; and the idea that fungi are a poor type of plant has been almost impossible to dislodge.

One of the principles of the *scala naturae* is that of plenitude. This means that there are no gaps in nature: there must, and will be, graded intermediates between any two organisms. The shrew, for example, might be the intermediate between a mouse and a mole; yarrow an intermediate between fennel and meadowsweet. If these examples seem rather poor, it is because the whole idea is nonsense and good examples hard to come by. Aristotle's successors had the same problem, but they did try hard and assumed that any embarrassing gaps would be filled eventually once the entire world had been explored.*

---

* In fact, it could be argued that the *scala naturae* is not as unreasonable as it sounds. Evolutionary descent with modification has, since its commencement, produced sufficient intermediate species to satisfy the principle of plenitude.

Aristotelian eponyms

Many scientific names derive from the common Greek names for the same, or similar, creature, so although some of the names Aristotle used make an appearance in modern nomenclature, it can hardly be said that he coined them. There are, however, several dedicated to him. It is sometimes thought that the genus *Aristolochia* (birthwort) is named for Aristotle, but the name derives from *Aristos*, the Greek word for 'best', presumably as in 'best at killing people' (the plant is responsible for more deaths than almost any other because of its use in more-than-usually-deadly herbal 'remedies').[54] A genus that is believed to be named for him is *Aristotella*, although this South American shrub hardly seems a sufficient or appropriate honour. Much better is *Silurus aristotelis* (Aristotle's catfish), a thoroughly Greek fish living in Greek rivers. I doubt, though, if Aristotle would feel quite so pleased with *Phalacrocorax aristotelis*, the common shag.

Theophrastus (circa 371–287BC)

Although honoured in a few scientific names, such as *Phoenix theophrasti* (Cretan date palm), Theophrastus is not a name with which many people are familiar. As Anna Pavord says in her excellent and detailed account of pre-Linnaean taxonomy, 'you can search Athens in vain for a memorial to this great man'.[55] Yet he is the undisputed 'father of botany' (a title conferred by Linnaeus himself). His two (surviving) books on plants – *Historia Plantarum* (*Enquiry into Plants*) and *De Causis Plantarum* (*On the Causes of Plants*) – mirror Aristotle's works on animals.

Theophrastus was born on Lesbos and studied first under Plato and then under Aristotle, providing him with the best academic CV in history. He was a close companion to Aristotle, his elder by fifteen years, accompanying him to Lesbos when Aristotle fled there. In 323BC he took over at the Lyceum when Aristotle had yet again to flee, this time to his native Chalcidice. Theophrastus continued his work there for a further, and remarkable, thirty-five years until his death at the age of eighty-four.

His works on plants were informed by close examination of specimens collected by him or for him. Many of these were provided by Alexander the Great, who, while on his business trips to Asia, ensured that interesting specimens were sent to Theophrastus at the Lyceum. Some of Theophrastus's descriptions seem rather fanciful, but his unlikely 'wool-bearing tree' makes sense once one realises he is talking about cotton. He shared the same practical preoccupation as his herbalist predecessors, with cultivated plants forming eighty per cent of the species in his books. It is clear from his works that he listened to the tales of gardeners (a time-devouring activity), writing about where the best place to grow vines might be and the tribulations of taking cuttings. He might better be described as the 'father of gardening', although Adam must take precedence. But to call him a mere gardener (my apologies to gardeners, I mean no offence) is to do him a disservice, for his influence on the science of botany, the science he founded, has been profound.

Theophrastus was not, nor ever intended to be, a great taxonomist, but as did Aristotle before him, he felt he had to make some arrangement of the organisms he was discussing. He wrote frequently of groups (kinds) of plants, such as when observing how seeds are held in various plants: 'In a vessel are those of the poppy and plants of the poppy kind . . . while many pot-herbs have their seeds naked, as dill, coriander, anise, cumin, fennel and many others'. The latter plants are all, of course, in the carrot family.

Although he discussed at some length how plants may be classified in a broader sense, his suggestions were all over the place. Like Aristotle, he used different categories for the same organism – sometimes habitat, sometimes whether a plant was deciduous or not, and so on. He also had a (tentative) over-arching classification, dividing the plants into four categories: tree, shrub, undershrub and herb. Here was the *scala naturae* taking full flight in botany for the first time – trees are bigger and more impressive than shrubs, which are bigger and more impressive than herbs. Not surprisingly, soon after establishing these groupings, he pretty well threw in the towel by admitting that some trees can grow like shrubs, some shrubs can grow like trees and some herbs can grow like undershrubs. Mallow, he wrote, 'grows tall and becomes like a tree'. No doubt he was mixing several *Malva* species into one, the tree mallow (*Malva arborea*), as we now know, being a distinct species.

The tree/shrub differentiation plagues us to this day. One man's shrub is another man's tree. Size is one of the poorest properties on which to base a classification. To take an example from the world of fungi, the morel (*Morchella esculenta*) is a substantial, and tasty, mushroom. Yeast is a single-celled organism just a few microns across. Yet they are much more closely related to each other than the morel is to, say, that other gastronomic joy, the penny bun mushroom (*Boletus edulis*), which in turn is closer to microscopic rust fungi. Theophrastus did not know what to do with fungi at all, but size considerations are much the same with the plants, if not so extreme.

The Fabaceae (the pea or bean family), for example, is the third-largest family of land plants after orchids and daisies. It has representatives in all four of Theophrastus's divisions – from the tiny black meddick (*Medicago lupulina*) to the ten-metre-high Judas tree (*Cercis siliquastrum*), both species with which Theophrastus would have been familiar. In fact, the largest of all the 'peas' is also one of the largest trees on the planet – the eighty-metre-tall tropical rainforest species tualang (*Koompassia excelsa*). All these plants share certain character-istics, including the structure of their flowers (typically sweet-pea shaped), fruit (pea pods) and their pinnate or tri-foliate leaves – although having all or any of these characters does not necessarily indicate that they belong to the same family, it just means they are likely to. Theophrastus, although he wrote often of leguminous garden plants, such as chick peas, beans and lupins, did not appear to appreciate, for example, that the acacia is a 'pea' too. But then neither did anyone else, for nearly two thousand years after his death.

Like his tutor, Theophrastus was a great scientist in a world that had not seen scientists before. His works on plants are extraor-dinary in their detailed descriptions of plant ecology, growth and structure. He formed no serious taxonomic arrangement of plants himself, but he did describe many of the characteristics – the differences – that would later form the basis of such a classifi-cation. Theophrastus was the first 'ecologist', noting the effects of soil type and climate on growth and species distribution. He showed some understanding of the two different branches of flowering plants: monocotyledons, whose seeds germinate with a single leaf, and dicotyledons, which have seeds that form two leaves. In *On the Causes of Plants* he wrote: 'the two, root and stem (of cereals), form a single continuous whole . . . All the leguminous plants have plainly two lobes and are double'. Unfortunately, he did not suggest this as a basis for any classification. It would have made a very good start had he done so.

Another differentiation within the flowering plants is the arrangement and positioning of the ovaries. They can be epigynous, hypogynous or perigynous – respectively above, below and around the female parts. These observations can only come from a patient study of the flowers in question, and Theophrastus was not surpassed in his assessments until the Renaissance. He also noted, no doubt from talking to growers, that many plants are dioecious (requiring both male and female plants to produce viable offspring, the word charmingly derived from the Greek for 'two households'): 'With dates it is helpful to bring the male to female; for it is the male which causes the fruit to persist and ripen'.

This was a modest man, who took the time to listen to others who knew the natural world and to study it, with care, himself. Perhaps this is his greatest gift. His plant descriptions are vibrant and colourful: this tree from the Carman region of Persia will be immediately identifiable to anyone who has seen one: 'These trees are all eaten away up to the middle by the sea and are held up by their roots, so that they look like a cuttlefish'. The mangrove, of course.

The great taxonomist William Stearn, in his book *Botanical Latin*, quoted Theophrastus's description of the European hop hornbeam (*Ostrya carpinofolia*):

> *It is similar to the beech in growth and bark; the leaves*
> *are shaped like pears at the base but they are much longer,*
> *narrowed to a point, and larger, and have many fibres,*

*which stretch out like ribs from a central straight large
fibre and thick; moreover the leaves are wrinkled along
the fibres and have a finely incisive edge; the wood is
hard, colourless and white; the fruit is small, oblong and
yellow like barley; it has shallow roots; it loves water and
is found in ravines* (translation by A. Hort, 1916).[56]

This could easily come out of a modern guide to trees, and Stearn compared it with a twentieth-century description of the same species. The two are virtually identical, except that the recent version includes dimensions. Theophrastus was a master at the art of analogy – a skill that has been inherited by and formalised in modern descriptive terminology. Where Theophrastus's leaves are 'shaped like pears', a modern botanist would say 'pyriform'; if our modern botanist wanted to say a leaf was 'shaped like pears but narrowed to a point', he might say 'pyriform, acuminate at apex', from the Latin *acuminare*, 'to sharpen'.

However, our topic is not just descriptions and taxonomy. It is also names, and to Theophrastus we owe many. Indeed most of the plant names employed for the next seventeen hundred years came ultimately from his pen.[57] Like Aristotle, he used the common names already available to him. Some already had Latin equivalents, so, for example, his *ptelea* and *batrachium* became *Ulmus* (elms) and *Ranunculus* (buttercups). But most of his names had no Latin equivalent and come down to us more or less intact as transliterations into Latin: *Aster, Antirrhinum, Cichorium* (chicory), *Anemone, Delphinium, Eryngium* (sea-hollies), *Hippophae* (sea-buckthorn), *Melilotus, Narcissus, Oenanthe* (water-drop-worts), Orchidaceae (the orchid family), *Origanum, Paeonia,* Poales (the grass order), *Raphanus* (radishes) and many more.[58]

Impatient of himself and of the time allotted to mortal man, Theophrastus frequently wrote that more research was needed – a plea not unknown in today's scientific papers. However, upon his death, with no immediate successor to carry on his work, the great and all too brief flowering of classical Greek biological enquiry came to an end. The works of Theophrastus were an important source for Roman scholars, but were subsequently lost to the West for a millennium, cared for by a more diligent Arab world. It was only during the Renaissance that his works become known anew, inspiring and informing a fresh, and long overdue, period of discovery.

Dioscorides (*circa* AD40–90)

The dominance of Greece gave way to that of a more pragmatic Rome, heralding a retreat to the utility of the remote past. However, despite their less-philosophical emphasis, it was the works of Roman writers that would inform Western thought on natural history, what there was of it, until the rediscovery of Aristotle and Theophrastus in the early thirteenth century.

De Materia Medica, the five-volume work of Pedanius Dioscorides, was the standard botanical text from the first to the sixteenth century. But as its name clearly indicates, it was not a true botanical work at all. It was a herbal, no different from many that had gone before, save for being considerably more comprehensive.

Dioscorides was, indeed, a medical man, and a Greek. His career began in the Roman army, where he served the health of some six or seven thousand men. Like all ancient herbalists, he would have collected many of his plants himself, and his observational skills are evident in precise descriptions of the growth habits of plants and their habitats. Six hundred plants are described, and their 'virtues' with them. The contributions of earlier botanists and herbalists are acknowledged; although he is mentioned by name only twice, Theophrastus is among them.

Despite being relied upon as the ultimate authority in botanical and medical matters for one and a half millennia, some of Dioscorides's descriptions tended to cause problems on account of their brevity – there is often little superficial difference between a plant that can cure and one that can kill. Illustrations that were added much later by others may or may not have been accurate, although after they had been through the hands of several copyists they certainly would not have been. And, although some of his plants were gathered from outside the area, his was a Mediterranean flora, ill-suited to northern parts. Efforts were made to find plants with similar properties native to other lands, but disaster awaited the patients of reckless herbalists.

Dioscorides had no interest in ordering plants, even alphabet-ically, although he did group together some related plants, such as the Apiaceae (umbellifers). The contribution for which we can be most grateful – in addition to his achievement in keeping the botanical torch burning, albeit dimly, for one and a half thousand years – is the large number of names he bequeathed us. In *A Natural Arrangement of British Plants* of 1821, Samuel Frederick Grays listed around one hundred names that originated with Dioscorides, albeit from early Latin transliterations of his Greek original, which has been lost. Their early origin has generally conferred high taxonomic status on Dioscoridean names, for example Asparagales, an entire order, which, rather obviously, includes asparagus. Family names include Cannabaceae, Hypericaceae (St John's-wort), Polemoniaceae (phlox) and

Polygonaceae (rhubarb, sorrel), while among Dioscoridean generic names are *Alyssum*, *Arum*, *Buglossoides* (in the borage family), *Clematis*, *Daucus* (carrots), *Galeopsis* (hemp-nettles), *Galium* (bedstraws), *Geranium*, *Myosotis* (forget-me-nots), *Petroselinum* (parsleys) and *Symphytum* (comfreys). Most of these names did not begin their lives as names for such high-level taxa; they were usually the names of single plants, and one at least remains so: *Myrrhis*, represented by its single species, *Myrrhis odorata* (sweet cicely). A few of his names appear only in the specific epithet. *Chamaedrys* was Dioscorides's name for germander speedwell; it is now *Veronica chamaedrys*.

Named for him, rather than by him, is the yam family, Dioscoreaceae. This is a particularly appropriate honour because, being a herbalist, he was much preoccupied with one of the main preoccupations of his patients – preventing the patter of tiny feet becoming too frequent. Dioscorin, the precursor of the synthetic hormones used in oral contraceptives, was isolated in *Dioscorea villosa*, the wild yam.

Pliny the Elder (AD23–79)

My grandfather was particularly fond of a strange book called *Enquire Within*, a thoroughly Victorian book of homely advice published from 1856 to 1976. I have his copy on my own desk now. He insisted that it contained the answer to any question I might have. Provided I want to know 'A Positive Cure for Corns', the 'Food of Blackbirds' or, more puzzling, the 'Use of an Elastic Fruit', *Enquire Within* is the place to look. However, on the matters of 'Remedies for the Diseases of the Eyes', 'The Method of Conveying Water' or 'The Food of Bees', it is silent. For these three I must refer you to the *Natural History* of Pliny the Elder.

Pliny was born in northern Italy to the equestrian class of Roman society – neither too grand nor too lowly. He served with the army as a doctor before embarking on a varied career in politics. Historians and other academics have mixed feelings about Pliny. He is often dismissed as 'nothing more than a mere compiler'[59] or, more cutting still, 'Pliny the Plagiarist'.[60] But there are few things that historians like more than compilers,* because they invariably quote from 'lost works', providing historians with endless, wistful fantasies. Without Pliny, many ancient writings would have passed into darkness: Cicero,[61] Varro,[62] Evantes,[63] Callimachus,[64] Crateuas and more owe a debt to Pliny. But Pliny also owes a debt to himself. Of all his many works – on the use of the javelin in the cavalry, rhetoric and grammar, a history of Rome – and more, only one remains: *Naturalis Historia* (*Natural History*). It is largely from within this book that we know of the others.

The encyclopaedic *Natural History* (AD70–77) was, without making any value judgements either way, the Wikipedia of its day and for many centuries after. In thirty-seven volumes it covers art, cosmology, meteorology, anthropology, geography, politics, agriculture, botany, zoology, medicine, mineralogy and many other subjects. My particular favourite is 'On the enormous price of fish'.†

Pliny estimated that his book contained twenty thousand facts, but modern scholars consider it to be near twice that number. He is not thought to have had first-hand knowledge of much of what he wrote, but he travelled widely and it is difficult to believe that he did

---

* He was commendably scrupulous in crediting his authorities, so 'compiler' is much more appropriate a description than 'plagiarist'.

† Pliny reported that one mullet cost eight thousand sestertii, which, by my calculations, is what a foot soldier in the Roman Army would earn in ten years. It was, as Pliny pointed out, cheaper to buy a cook.

not include many of his own observations in addition to those of nearly five hundred sources. He quoted both Aristotle and Theophrastus (in the latter case sometimes referring to lost works in zoology). Uncredited lines of Dioscorides are also found in *Natural History*, but this is more likely due to a common source than to Pliny helping himself to the works of his near-contemporary.

He was famously uncritical of his sources, reproducing many fantastical stories, although whether this was to amuse or inform I do not know:

> *. . . the fiercest animal is the unicorn, which in the rest of the body resembles a horse, but in the head a stag, in the feet an elephant, and in the tail a boar, and has a deep bellow, and a single black horn three feet long projecting from the middle of the forehead. They say that it is impossible to capture this animal alive . . .*

And:

> *In its neighbourhood (source of the Nile) there is an animal called the Catoblepas, in other respects of moderate size and inactive with the rest of its limbs, only with a very heavy head which it carries with difficulty – it is always hanging down to the ground; otherwise it is deadly to the human race, as all who see its eyes expire immediately . . .*

However, he was sensibly suspicious of Aristotle's salamander, finding its reputed ability to endure fire highly implausible.

Natural History is largely a practical book. Indeed, in his introduction, Pliny expressed the hope that it would be used by farmers and artisans and not just for the amusement of the educated. Much is given over to gardening and agriculture, and even more to the uses of plants as medicine. Its zoology is extensive but observational in nature, with little in the way of a classification save such things as classifying birds by their feet: taloned, clawed or webbed. Botanical classifications, such as they are, are of the 'Ten varieties of Linden-tree' and 'Beet – four varieties thereof' kind. We owe many generic names to Pliny, such as *Mercurialis*, *Euphorbia*, *Convolvulus*, *Solanum* and *Viola*, although his scholarly failings and sometimes careless translations from Greek have suffered much criticism. Some of it is unfounded. In particular Pliny was right about shrews. His *Sorex* was replaced with *Mus araneus* ('spider mouse') by one sixteenth-century busybody (see p. 18), not realising that *Sorex* was a perfectly good Latin name for a shrew.

Pliny's work, like that of Dioscorides, became an essential source of knowledge to the many generations that followed, for whom appeal to authority rather than scientific investigation was to become the only path. In return for his efforts Pliny was immortalised in the genus *Plinia* (in the myrtle family) by Charles Plumier in the eighteenth century.

Not until the sixteenth century were there any real, lasting advances in the naming of the living world. Indeed, without the search for new knowledge, that which existed became lost or corrupted. The thoughtful works of Aristotle and Theophrastus were still languishing in the East, and the handful of classical herbals that were available through those long years were copied and copied and copied. With all this copying and no one to check for accuracy or even sense came errors, and little more than Chinese whispers remained.

The bestiaries

In zoology it was no better. Animals (at least animals that you couldn't eat) were of interest primarily for their Christian symbolism. Thus were the bestiaries created. In these beautiful tomes, the fashionable coffee-table books of the twelfth century, tales of mythical and real animals were mixed indiscriminately. The partridge and phoenix, halcyon and hare, for example, are illustrated and discussed with never a mention that only two of them actually exist. Despite the moral tone that emanates from every page, there is nevertheless some reasonably good zoology. The descriptions (even of the animals that did not exist) are brief but clear: many hundreds of years before the phenomenon had become accepted, the fact that birds such as the swallow and the stork migrate is stated matter-of-factly in these books. Several taxonomic names have been derived from these fabled creatures: *Dragonana* (unimpressively a genus of leafhoppers), *Sphinx* (a genus of moths), *Basiliscus* (a genus of lizards) and also Amphivia, a legendary species of fish that was as happy to run around as it was to swim.

Albertus Magnus (circa 1204–1280)

Almost the only botanical light to illuminate the Middle Ages was that of Saint Albertus Magnus, a Dominican bishop from Germany. The works of Aristotle and other classical writers had been rediscovered[65] and he set to work examining them.* Among them was a treatise on plants, which he thought to be by Aristotle

---

* 'Lost' manuscripts in Greek or Arabic existed, mostly in the Middle East. They were 'recovered' by being translated into the much more accessible Latin in the twelfth and thirteenth centuries.

but was in fact, via a tortuous chain of copying and translation, found to be derived from Theophrastus. On this he based his own book, *De vegetabilis*.

Magnus's basic taxonomy was the same as that of Theophrastus – trees, shrubs, undershrubs, herbs – and he ran immediately into the same problems. Biennial plants, in particular, were troublesome, because they are herb-like in their first year and tree-like, or at least woody, in their second. Not surprisingly, Magnus had a lot of problems with sex – he didn't consider that plants indulged in such unseemly activities. He nevertheless ascribed male and female status to plants according to their general character: thin/hard/dry for male and broad/soft/moist for female.[66] Some of his plant physiology was highly suspect, too: he thought that thorns were exuded from plants by virtue of viscosity, then drawn out and sharpened by the warmth of the sun, only to be bent backwards by that viscosity.[67]

Although he relied on Theophrastus, Magnus shows every sign of having examined plants himself. He noticed details of flower structure, which are still important today, the difference between non-vascular plants (mosses, liverworts) and vascular plants (most of the rest), and the division between dicotyledonous and monocotyledonous plants. Unfortunately, like Theophrastus, he had no immediate successors, so botanical study stayed on hold for another two centuries.

Albertus Magnus's name is much honoured today in the names of colleges and schools and also in the unusual 'double eponym' *Alberta magna*, the Natal flame bush. Magnus has been lucky in this choice. *Alberta magna* is the type species of its genus (see p. 118), so when it was found that the genus should be split into two, the plant remained in the established genus. If another plant had been the type species, he would have been remembered in the considerably less euphonious *Razafimandimbisonia magnus*.

Brunfels (1488–1534), Bock (1498–1554) & Fuchs (1501–1566)

In parallel with the religious, political and artistic revolutions of the Renaissance, science and its handmaiden philosophy flowered as never before. Authority gave way to observation, and a serious examination of the natural world began. The science of botany was developed by a succession of remarkable men, laying the groundwork for Linnaeus and his contemporaries in the eighteenth century.

The standard contemporary work on British plants, *New Flora of the British Isles* by Clive Stace, is almost entirely without pictures, relying instead on careful and comprehensive descriptions. However, people have always liked pictures to help them with identifications. Without the modern specialised vocabulary of botany (even the word 'petal' is a relatively new coinage), it was practically impossible to describe a plant with any expectation that someone else would reliably know what you were talking about. Enter Otto Brunfels and his 1530 *Herbarium vivae icones* (*Living Images of Herbs*). There were, of course, pictures of plants available in the late Middle Ages, but they had undergone the same creeping corruption suffered by ancient texts and become stylised to the point of absurdity. Brunfels started looking at plants afresh and commissioned a highly talented wood engraver, Hans Weiditz, to produce illustrations of some two hundred and sixty plants.

The pictures are exquisite, showing all the details necessary for the identification of a species at little more than a glance. Brunfels combined them with descriptions taken from existing writings, correcting an error here and there and, unfortunately, adding a few as well. There are no taxonomic innovations, although he did change the genus names of some plants that had been indiscriminately lumped together because they happened to cure the same disease. But really it is the pictures that matter, as Brunfels himself recognised:

> In this whole work I have no other end in view than
> that of giving a prop for botany; to bring back life to
> a science almost extinct. And because this has seemed
> to me to be in no other way possible than by thrusting
> aside all the old herbals, and publishing new and really
> lifelike engravings, and along with them accurate
> descriptions extracted from ancient and trustworthy
> authors, I have attempted both; using the greatest care
> and pains that both should be faithfully done.[68]

Brunfels, together with Hieronymus Bock and Leonhart Fuchs, made up a blessed trio of central European botanists known as the 'German fathers of botany'. In his 1542 *Historia Stirpium* (*History of Plants*), Fuchs took the role of a 'super Brunfels', with better descriptions, better illustrations and more plants. The book covered a remarkable five hundred species, including one hundred domesticated plants, and heralded a flood of species (such as the newly discovered exotics maize and pumpkin), that would soon overwhelm botany.[69]

Hieronymous Bock (or Hieronymous Herbarius or Hieronymus Tragus – evidently he could never make up his mind)[70] – was unlucky in his personal life. He died of consumption in 1554 at the age of fifty-six, having outlived his younger wife and eight of his ten children. Professionally, however, he is considered to be one of the greats. His magnum opus was called, prosaically, *Kreuterbuch* (*Plant Book*), but it was quite revolutionary, setting the standard for botanicals to this day. Like our own Professor Stace, he did not include illustrations (although an illustrated 'director's cut' was published a few years later). He relied instead on careful descriptions taken from personal observations of more than eight hundred species. Unlike Brunfels and Fuchs, whose main objective was to provide an accessible herbal and who therefore settled for alphabetical order, Bock attempted a classification:

> *In describing these plants, I have followed the principle of joining in my book the plants that nature seems to have joined by a similitude in form. But I have separated each plant into its own chapter. I did not want to follow alphabetical order, as the old herbalist wont.*[71]

Not that he made a particularly great job of it. He was still hampered by the old notion that trees, shrubs and herbs were fundamentally different, or, more precisely, occupied different regions on the *scala naturae*. However, he did start to classify plants according to their flowers and their seeds, a method that has proved useful, if sometimes highly misleading, to this day.

The writings and illustrations of Brunfels, Fuchs and Bock mark the early beginnings of the modern age of botany and of taxonomy in general. By the time Brunfels was published, printing using moveable type was eighty years old, so all of their works were readily available to those with an interest and the where-withal to pursue it. Following their work, botanists (and, to a lesser extent, zoologists) with a zeal for classification came thick and fast, contributing during the next two hundred years to an ever more substantial body of knowledge. The flood of new species into Europe owed to the efforts of explorers certainly made a classification system worth having; essential, in fact.

The three German fathers of botany are remembered in modern generic names: Brunfels for a small genus of plants in the nightshade family, *Brunfelsia*, and Fuchs, of course, for *Fuchsia*. Bock, who is known in the botanical world chiefly under his Latinised name, Tragus,* is memorialised in the bur-grass genus *Tragus* and the spurge *Tragia*.

Cesalpino (1519–1603)

Andrea Cesalpino is known as another 'father', the 'father of Italian botany'. This honour was conferred by the publication in 1583 of his *De Plantis* (*On Plants*), which contains minutely detailed descriptions† of fifteen hundred plants. Cesalpino was a classicist and philosopher as well as a botanist, and there is some evidence to suggest that he attempted to impose at least some Aristotelian order on the living world.

He used Aristotle's esoteric ideas of 'essence', 'difference' and nested 'genera' and 'species'. The 'essence' (see p. 151) is the essential character of a species or genus. In practical terms, this

---

* It was fashionable at that time among men of learning to Latinise one's name. *Tragus* is the Latin equivalent of the German word *Boch* – 'he-goat'. It is often thought that Linnaeus Latinised his own name, but he inherited it 'pre-Latinised' from his father Nils Linnæus, who named himself after the large linden tree (*lind* in Swedish) in the family grounds. Carl was later effectively to 'de-Latinise' his name on his ennoblement to Carl von Linné, only to have this Latinised all over again as Carolus a Linné.

† Some of his descriptions, such as that for Aristolochia, fill two densely written pages.

meant choosing a character (or characters) that embodied the essence of a group or individual. If several organisms possessed a particular form of this character, they could be grouped together. However, it is possible to use different characters for different levels of classification: one thing (fruit, for example) might work at the level of the modern order, while another (perhaps flowers) may be best suited to the (modern) genus. Choosing such characters is notoriously difficult. Leaf shape, for example, might seem a likely contender because they come in a large number of varieties. Unfortunately they are extremely inconsistent within almost any group of flowering plants. For instance, the Chenopodiacae (goosefoots), a true natural group, has leaves ranging from large and broadly triangular through small and oval to tiny and cactus-like. Based on leaf shape, you could not necessarily say that a plant was a goosefoot, although if you already knew that it was, you could probably tell which one.

Cesalpino's contention that a single character at any level could be used reliably to determine plant groups is seldom true. But the character he settled on for what we would now consider a plant family was extremely well chosen. He chose fruit (nuts and seeds). Being so important in the life of a plant, these fitted well with any underlying philosophical considerations and had the great advantage of being extremely variable between what were presumed to be natural groups but reasonably consistent within them. However, he used a different character for his highest level of plant classification and, being a fan of Theophrastus, went for stems. In other words, he inherited a condensed *scala naturae* arrangement of tree and herb (his *Arboreae* and *Herbaceae*). Inevitably this caused conflict, but Cesalpino was a pragmatist and he employed a certain amount of selective blindness in order to ignore characters that indicated a classification that defied his well-practised instincts.

At the level of the modern family his arrangements are remarkably accurate. The families Fabaceae (beans and peas), Apiaceae (carrots), Brassicaceae (cabbages), Compositae (daisies) and Lamiaceae (mints and deadnettles) were of his devising and survive largely intact from his work.

In studying the structure of plants in great detail (and not just looking at stems and nuts), Cesalpino (who is commemorated in the legume subfamily *Caesalpinioideae*) was able to form the first systematic taxonomy ever devised. It was he who determined that morphology was the only basis for a classification. Linnaeus is often considered the father of taxonomy, but perhaps it is Cesalpino who truly deserves the title.

Bauhin (1560–1624)

Cesalpino's *De Plantis* is a dense and forbidding work, but that of his successor Gaspard Bauhin is a much more modern, cheerful-looking tome. *Pinax Theatri Botanici (Illustrated Exposition of Plants)* of 1623 is nicely laid out in chapters and sections, the sections denoting something similar to what we would now accept as families and, in what appears to be a creeping case of one-up-manship, including six thousand species. A few of the names start to look decidedly modern, with such as *Pinus sylvestris* (Scot's pine) and *Raphanus sativus* (garden radish) making an appearance.

It is sometimes said that Bauhin invented the binomial names that Linnaeus eventually adopted into his naming system. There is a little truth in this. Bauhin did sometimes employ binomials but, as was the common usage, most were not binomials as we know them. They were 'diagnostic' names, which first declare the genus and then state what is different about it. So, when Bauhin wrote of *Raphanus sativus*, he was both naming and describing the plant as 'radish, the cultivated one'. By *Raphanus aquaticus foliis in profundas lacinias divisis*, he meant 'radish, the one growing in water with deeply divided leaves'. A little like the episode titles of *Friends*, perhaps.

However, it was Bauhin who was largely responsible for species names being in these two parts: genus and specific epithet (even though the latter could consist of several words). That they should be of only two words was a rule established by Linnaeus, although not entirely his invention. The exception to this for Bauhin was in species that were lone representatives of their genera. *Larix* (larch) is just plain *Larix* because there were no other larches known at the time from which a differentiation would have been required.

Is Bauhin's classification any better than most of his predecessors? Not much. Some of the 'sections' that Bauhin used seem to make sense, but with no real idea of what characters to use to differentiate his groups, he ended up all over the place. For example, apples, pears, plums, cotoneaster and other members of the order Rosales rub shoulders in the same section as oranges and lemons, their very distant cousins in the order Sapindales. Less fruity trees are split between other sections, but we find the birch in with the elm, its third cousin once removed, but in a different section from its first cousin, the oak. Inevitably Bauhin had a lot of trouble with peas, lumping several species of clover together with other plants that have trifoliate leaves.

Neither Bauhin nor his contemporaries can be blamed for their shortcomings. The 'natural' classification for which they were beginning to search is extremely hard to find. Plant and animal physiologies, on which classifications can to a large extent be based, were in their infancy. More important still, the obvious fact that plants (and animals) fell into more or less clear groups never prompted the question of why. Why would God produce organisms that had things in common? Why not thousands of species with nothing much in common at all? No one knew; no one asked.

John Ray (1627–1705)

Even in the early Enlightenment, God was not required to give an account of himself. The great English naturalist John Ray described his own endeavours as being 'to illustrate the glory of God in the knowledge of the works of nature or creation'. The language is telling: 'illustrate', not 'explain'.

John Ray is sometimes known as (you guessed it) the 'father of English botany'. Although of humble origins – he was the son of a blacksmith – he went to Trinity College, Cambridge. Of his twenty-three published works, ten are botanical. His major work, *Historia Plantarum* (1686), is an extraordinary tome that purports to list all the species of plant known at the time – 18,655 of them.[72] If this seems like an awful lot of plants, it is because his idea of a species was a narrow one – he was a consummate 'splitter' (see p. 128).

Ray had spent several years exploring the flora of his native islands and had also ventured into Europe. Inevitably he accumulated a vast amount of knowledge about the plants he examined, which put him in an excellent position to make a sensible classification. He was not a medical man and neither was he wedded to the classical philosophical notions that had led so many astray.[73] He did embrace the tree/herb differentiation, but nevertheless his groupings largely reflect modern classifications. His general thesis was one of informed common sense using a combination of characters: 'The first condition of a natural method must be that it neither splits plant groups between which apparent natural similarities exist nor lumps such with natural distinctions'.[74]

For example, even though the leaves of the white deadnettle look remarkably like those of the stinging nettle (in the modern family Urticaceae), Ray correctly determined that their opposite alternate pairing on a square-section stem and their labiate flowers (the same shape as those of a snapdragon, but small and white) put the white deadnettle firmly in the mint family (the modern Lamiaceae). He also managed to spot two fundamental aspects of plants that had been noticed before but not given due prominence in classification: the difference between the monocotyledons and the dicotyledons[75] – words he coined – and the distinction between the angiosperms and the gymnosperms – literally 'shelled seed' (flowering plants) and 'naked seed' (conifers, ginkgo and others).

If you think I have been ignoring zoology in my brief history, that is because it was largely ignored by most, if not all, early naturalists. John Ray changed this. Not content with his trans-forming contributions to botany, he also turned his remarkable observational skills on the animal kingdom. His classification of tetrapods and birds split them dichotomously: cold-blooded with only one heart ventricle (reptiles and amphibians) and warm-blooded with a double heart ventricle (everything else). The 'everything else' was split into live-bearing and egg-bearing, that is, mammals and birds. He also managed to answer the question that exercises every child to this day: are whales and dolphins fish? No, they are not, he concluded, they are what Linnaeus would later name mammals.

He went into more details about birds, producing the first of what would become a large publishing genre – a guide to birds. *Ornithologia* was published under the name of his late friend and patron Francis Willughby, being in part based on his notes. Ray and Willughby carefully examined and dissected their subjects, forming a reasonably clear idea of which bird is related to which. Feet, Ray decided, were the most important attribute for deciding a classification, and his arrangements are suprisingly modern. Oddities occurred, of course: he placed the kingfisher with the woodpecker, although they are now known to be in completely different orders. It is indicative of the times that Ray felt obliged to consign phoenixes, griffins, harpies and ruhk to the wonderland in which they had always belonged. But this otherwise admirable incredulity let him down when he rejected the condor. He refused to believe that any bird could get that big.[76]

What little classification had been done in either the plant or animal kingdoms up to the time of Ray is indicated in the low number of accepted rankings he used. With no firm concept of a larger hierarchical classification, seventeenth-century biology had to make do with a vague and meagre three: kingdom, genus and species, with the notion of 'genus' a rather moveable feast. However, Ray improved on this, albeit informally, by suggesting the idea of higher 'genera', which would encompass lower 'genera'.

Although Ray used many binomial names, he was a sucker for really long ones: *Equisetum palustre brevioribus foliis polyspermon* ('horsetail with short leaves, many seeds and growing in marshy areas') gives an idea.[77] I rather fancy that he was compensating for his own, short, disyllabic name (I have some fellow feeling here) and jealous of all those continental biologists who could revel in glorious polysyllables. In fact, he eventually followed fashion by Latinising his own name to Joannis Raii, tripling the number of syllables at a stroke. He is remembered in the names of several species, *Polygonum raii* (Ray's knotweed) being one of them. There was once *Rama raii*, the Atlantic pomfret, until someone changed its name to *Brama brama*. It's a miserable-looking fish anyway, so perhaps Ray is better off without it.

Joseph Pitton de Tournefort (1656–1708)

John Ray brought some semblance of order to what was becoming an increasingly chaotic body of knowledge. Not everyone took notice of him, however. Joseph Pitton de Tournefort, whose most important work was *Élémens de botanique* of 1694, made an important contribution to taxonomy in that he formalised the concept of 'genus', an idea that had been buzzing around for centuries, landing sometimes here and sometimes there. He defined it simply as a 'cluster of species', which 'bring together those plants which resemble one another, and to separate them from those which they do not resemble'.[78] He also suspected the existence of higher taxa, which he called 'classes'.

His goal, he said, was to 'reduce each species to its true genus'. He was reasonably successful, but his placing of genera within a larger hierarchy was extremely hit-and-miss. Ignoring Ray's catholic approach of accepting a variety of characters to determine taxonomic position, he stuck instead to flowers and fruit. Mostly it was flowers, and even with them it was largely the corolla (petals that make up a blossom) that interested him. This immediately caused problems, as is clear from the diagram of Tournefort's family tree later provided by Linnaeus. For example, bell-shaped corollas abound in the flowering plants, so Tournefort grouped such things as deadly nightshade (*Atropa belladonna*) with solomon's seal (*Polygonatum commutatum*) and lily of the valley (*Convallaria majalis*). The second two are monocots and therefore distant from deadly nightshade, as Ray could have told him. Worse still (although we should not be surprised by now) is that his primary division was between trees and herbs. These failings aside, Tournefort's contribution to taxonomy was immense and much of his system, for good and ill, was taken up by Linnaeus fifty years later.

Augustus Quirinus Rivinus (1652–1723)

I have described only a handful of the naturalists who have advanced (or impeded) the cause of taxonomy. One more deserves a mention. Augustus Quirinus Rivinus, a German physician and botanist, published several works on botany, including *Introductio generalis in rem herbariam* of 1690. He made three important innovations. Prior to Tournefort he went some way toward formalising genera by allowing them one consistent name each. Secondly, no doubt despairing of interminably long names, he suggested that everyone might be better off with just two. This (extremely good) idea received only intermittent acceptance until Linnaeus made it a rule (as only Linnaeus could have done). Thirdly (and it is time to open the champagne), he thought that plants should be classified according to their flowers and other detailed characteristics, irrespective of how big the plants themselves were. At long last the intellectual constipation that had afflicted botany for two thousand years was purged.[79] The way was now clear for a great mind, a great organiser, a great clerk, to put the world in order.

Linnaea borealis

CHAPTER VII

The FATHER of TAXONOMY

T HERE IS A TOUCH OF THE MESSIANIC IN MANY OF THE STORIES ABOUT Carl Linnaeus (1707–1778). Like the twelve-year-old Jesus of Nazareth, he was a precocious child; Linnaeus is said to have loved flowers and played with them as a small child, quickly learning their names. At the age of eight, he even planted a small garden of his own: 'Carl's garden', as it was known. In the early part of his career he spent time in the wilderness, not in Judea, but in Lapland. He had many followers, some even known as 'Apostles', who went all over the world to collect plants for him (and often came to a sticky end, as did many of Jesus's followers). Linnaeus also had his own Pharisees to mock him, but was destined to receive the unstinting praise of all – well, nearly all.

Many great things were said about this great man. 'With the exception of Shakespeare and Spinoza, I know no one among those no longer living who has influenced me more strongly!' Goethe wrote. Jean-Jacques Rousseau, who seldom had a good word to say about anybody, sent a message to Linnaeus: 'Tell him I know of no greater man on earth'.[80] Cuvier called Linnaeus the 'Pliny of the North'.[81]

We are also informed that he 'fundamentally reorganised the whole field of natural history, raising it to the height it has now attained', that one of his books was 'the greatest achievement in the realm of science' and another 'a masterpiece' that 'no one can read too often or admire too much'.[82]

These quotations are, unfortunately, from the pen of Linnaeus himself. Yet despite this outrageous self-congratulation, it is difficult to think of Linnaeus as anything other than a modest man. He started his life simply enough, as the son of a curate of adequate means. Carl Linnaeus was a Swede – Sweden's most important contribution to world culture until Abba. He was born in Råshult in the far south of the country, at the beginning of a century that was to be the golden age of botany.

Carl's grammar-school headmaster was intrigued by his failing student's interest in botany and sent him to Dr Johan Rothman, a senior master at Växjö high school. It cannot be said that Linnaeus flourished here either. The curriculum, designed to produce clergymen and officials,[83] was a dry and dusty diet of Latin,* theology, Greek and mathematics. Lessons started at a mind-numbing six in the morning, after prayers, continuing until 5 pm. Rothman assured Linnaeus's distraught clergyman father that, while his son would never be a priest, he could at least be a doctor. Rothman personally taught the young Carl the fundamentals of medicine and also botany, including the existing systems of plant classification.

Linnaeus owes much of his success to the people he met. However, this was no 'old boys' network'; the wealthy, the well-connected and the intellectuals of his day sought him out for his genius alone. Natural history was highly fashionable at the time, with huge collections of novelties being amassed by the

---

* The Latin at least gave him the keys to the kingdom of academia, for without it he would have been excluded from the learned world of Europe.

well-to-do as conversation pieces. The exotic plants and animals that were flooding into Europe from newly discovered territories required someone skilled in providing them with a context. Linnaeus was such a person.

Unfortunately, he was prone to making a bad first impression, his brash self-confidence being too much for many people. Nevertheless, at Lund, a down-at-heel university not too far from Linnaeus's home, he made a good first impression on an eminent doctor and naturalist, Dr Kilian Stobæus. As was to be a pattern over the next few years, the good doctor took the impoverished Linnaeus into his home, lent him books and permitted him to attend lectures for free.

During his short time with Stobæus, he explored the flora and fauna around Lund, expanding his knowledge of plants and animals. It was on one of these trips that he first encountered the remarkable *Furia infernalis* (see p. 28). Linnaeus learned much from his mentor, including the very practical matters of how to pin insects and preserve plants in a herbarium. The internationally important Linnaean* Herbarium and collections, which are protected for study and posterity within a vault, were begun under the tutelage of Stobæus.†

After nine or ten months, in May 1728, Linnaeus returned home to recover from an illness. He then made his way to another university that had seen better days, Uppsala, to the north of Stockholm, where he was to become, by a long way, its most

---

* The contrary spelling used by the Linnean Society in its own name is due to its being derived from Linnaeus's noble name, Carl von Linné.

† Dr Stobæus is remembered in *Stobaea*, a rather prickly-looking genus of plants.

famous alumnus. Here, in 1729, he met Peter Artedi, another student and a keen naturalist. The two young men became very close. Their joint interest led them to a pact: to divide between them the entire living world and study and classify it for the benefit of mankind. Among the animals, Artedi took amphibians, reptiles and his chief love, fish. Linnaeus took birds and insects. The mammals they divided between them. Artedi had a particular fascination with the carrot family, but Linnaeus took the rest of the plants. Evidently they thought this would not be enough to keep them occupied, so they determined to tackle minerals as well. Their pact came with a solemn oath to complete the task should one of them die before it was completed.

It would fall to Linnaeus to fulfil this pledge a mere eight years later, when Artedi fell into a canal in Amsterdam (a really bad place in which to stagger home after a good night out with friends).

> *Here lies poor Artedi, in foreign land pyx'd*
> *Not a man nor a fish, but something betwixt,*
> *Not a man, for his life among fishes he past,*
> *Not a fish, for he perished by water at last.*[84]
> [A 'pyx' is a type of box.]

Epitaphs were often less sensitive in the eighteenth century than would be tolerated today. This one by the botanist Olof Celsius (of whom more shortly) was discovered (in Latin) on the back of Linnaeus's copy of Artedi's *Ichthyologia* (*On Fish*), which was published in 1738.* Having eventually retrieved his friend's papers

---

* A less frivolous memorial was provided by Linnaeus, who named a member of the carrot family (Apiaceae) after his friend; it is a genus with just one species – *Artedia squamata*.

and collections from a rather put-upon landlord with the help of a wealthy friend, George Clifford (see below), Linnaeus edited Artedi's work and ensured that it was published.

Linnaeus first encountered Olof Celsius in 1729, not long after his first meeting with Artedi. He was a professor at Uppsala (and the uncle of Anders Celsius, the inventor of the centigrade scale of temperature measurement).* Linnaeus met him by chance one day in the botanic garden, and impressed him by naming all the plants around them and even placing them within the system laid down by Tournefort. Upon discovering that Linnaeus had a collection of six hundred species of wild flowers, Celsius duly invited him back to his house, eventually providing him with accommodation and the run of his library. Linnaeus was practically destitute at the time so the offer was welcomed with as much relief as excitement.

Later that year, breaking with the student tradition of addressing a poem to one's professor, Linnaeus presented Celsius with a dissertation entitled *Praeludia Sponsaliorum Plantarum* (*Prelude to the Betrothal of Plants*). An early exploration of a subject that Linnaeus would return to later – sexual reproduction in plants – the paper so delighted Celsius that he showed it to Olof Rudbeck, a professor of some influence at the university. In his turn, Rudbeck gave board and lodgings to the still impoverished Linnaeus, although how he found room for him, I do not know – Rudbeck managed to get through three wives, fathering

---

* Anders Celsius, who obviously hadn't thought things through properly, placed the boiling point of water at zero and the freezing point at one hundred; the whole thing was upside down. It was our own Linnaeus who turned things around – without him we would all evaporate when our thermometers showed 0 degrees C. Maybe a campaign for degrees Linné is in order.)

twenty-four children in the process. He also gave Linnaeus his first job, presenting what turned out to be extremely popular lectures on botany to his fellow students. Instead of the normal seventy to eighty students he would sometimes have the attention of three or four hundred.

Rudbeck is commemorated in the American genus *Rudbeckia*, of which Linnaeus wrote, in his typical purple prose: 'I have chosen a noble plant in order to recall your merits and the services you have rendered, a tall one to give an idea of your stature; and I wanted it to be one which branched and which flowered and fruited freely, to show that you cultivated not only the sciences but also the humanities. Its rayed flowers will bear witness that you shone among savants like the sun among the stars'.[85] Appropriately enough, *Rudbeckia* sports about two dozen accepted species, nicely matching the number of the professor's children.

Linnaeus's big adventure

Linnaeus is sometimes characterised as a dry figure – at least two of his most influential works read a little like a computer spread-sheet – but in 1732 he made a journey that was full of romance. He approached the Royal Society of Sciences at Uppsala, suggesting that it fund a one-man expedition to Lapland by someone know-ledgeable in plants, animals and minerals. The person should also be young, fit, single and most certainly Swedish. Someone, perhaps, like Linnaeus. He kept a journal for his own use, which was not published until after his death.* Much of it is of the familiar 'Continuing my journey at sunrise, I saw some sepulchral

---

* By the founder of the Linnean Society, Sir James Edward Smith, in an abridged and translated version in 1811.

mounds near the church of Jattedahl' variety of traveller's notes. But this was no ordinary traveller, for he recorded nearly every plant, lichen, fungus, animal and mineral he encountered. The journal betrays his dissatisfaction with existing systems of classification, notably Tournefort's system, berating his predecessor for inaccuracies twice in the first eight pages. On one occasion, evidently having been on his own for far too long, he was inspired to name a genus there and then:

> *Chamœdaphne of Buxbaum was at this time in its highest beauty, decorating the marshy grounds in a most agreeable manner. The flowers are quite blood-red before they expand, but when full-grown the corolla is of a flesh-colour. Scarcely any painter's art can so happily imitate the beauty of a fine female complexion; still less could any artificial colour upon the face itself bear a comparison with this lovely blossom. As I contemplated it I could not help thinking of Andromeda as described by the poets . . . Andromeda is represented by them as a virgin of most exquisite and unrivalled charms; but these charms remain in perfection only so long as she retains her virgin purity . . . Hence, as this plant forms a new genus, I have chosen for it the name of Andromeda.*[86]

Linnaeus drew a (frankly) crude sketch of Andromeda chained to her rock, beset by a dragon, alongside *Andromeda* nestling in some mud, beset by a newt. The drawing is not exactly Edward Poynter, but one or two details of Andromeda get thoughtful attention. The plant is now known as *Andromeda polifolia* (bog rosemary).

Another plant he came across was the twinflower, *Campanula serpyllifolia*, 'whose trailing and verdant leaves were interwoven with those of Ivy'. The plant, his favourite, was later renamed *Linnaea borealis* by the botanist Johan Gronovius, as Linnaeus much later explained in a particularly unconvincing display of modesty: '[It] is a plant of Lapland, lowly, insignificant, disregarded, flowering but for a brief space – from Linnaeus who resembles it.'

Linnaeus was not concerned simply with plants. He gave a great deal of attention to animals and birds, too (although as one unfortunate Northern hawk-owl found to its cost, he was not averse to the now less respectable technique of blasting one out of the sky with a shotgun in order to make an accurate identification). Near Lycksele, he encountered a peasant who had shot a beaver. The man explained that the locals sold the skins and boiled the animals for food, eating only the feet and throwing the rest away – there presumably being more meat on a beaver's foot than one might guess. Upon being told that beavers eat the bark of birch, aspen and pine, Linnaeus went on to 'wonder no naturalist has classed this animal with the Mouse tribe'. Later he was to do just this, placing both the mouse and the beaver within his order Glires.*

Linnaeus did not always get things right: sharing his Glires with mice and beavers are rabbits and hares, which belong to a different order. This does not seem too egregious an error, but he also included the more difficult-to-excuse rhinoceros. Perhaps the rhinoceros has teeth not unlike that of a mouse (only bigger, obviously), for later in his journey Linnaeus observed:

---

* Modern taxonomy has little of fundamental difference to say on the matter, as they are still in the same order, now called Rodentia.

Close to the road hung the under jaw of a Horse, having six
fore teeth, much worn and blunted, two canine teeth, and
at a distance from the latter twelve grinders, six on each
side. If I knew how many teeth and of what peculiar form,
as well as how many udders, and where situated, each
animal has, I should perhaps be able to contrive a most
natural methodological arrangement of quadrupeds . . .

In his *Systema Naturae* (see p. 191), Linnaeus would use such a
system to arrange the mammals, and this is the first time he
mentions the notion. For the most part it works surprisingly well,
although, as we have seen, things could go awry on occasions. He
did not drop the udder idea either.

At least part of his aim was to learn something about the way the
Lapps lived. He had great respect for the people and their mode of
life, pointing out how happy they were, despite the hardship of
their lives, and how healthy they were, despite their unrelenting
diet of reindeer and reindeer milk and a laissez-faire view of
personal hygiene (Linnaeus's enjoyment of an otherwise
appealing dinner was ruined after he saw how they cleaned the
cutlery with saliva). As well as general-purpose anthropology,
Linnaeus also engaged in ethnozoology, ethnobotany and ethno-
mycology (a taste of which I cannot resist):

I was also shown the Agaric of the Willow which has
a very fragrant scent. The people have assured me it
was formerly the fashion for young men, when going to
visit their mistresses, to use this fungus as a perfume,
in order to render themselves more agreeable . . .

In his *Flora Lapponica*, which was to be published five years later, he expanded on this encounter:

> *O whimsical Venus! in other regions you must be*
> *treated with coffee and chocolate, preserves and*
> *sweetmeats, wines and dainties, jewels and pearls,*
> *gold and silver, silks and cosmetics, balls and*
> *assemblies, music and theatrical exhibitions: here*
> *you are satisfied with a little withered fungus . . .*

He named this magical fungus *Boletus suaveolens*, now *Trametes suaveolens* (the specific epithet means 'sweet-smelling'), a pale bracket fungus that grows on willows, with an almond/anise scent. It is rare, in the UK at least, so we may have to stick to a decent aftershave.

Linnaeus's big adventure became central to his character and his reputation. In the most famous painting of him he is dressed in traditional Lapp costume, holding one of their ritual drums. While not on a par with other expeditions of the time, on which antagonistic natives, deadly animals and disease were constant threats, his journey to the North nevertheless confirmed Linnaeus as being someone prepared to fearlessly wrestle with nature first hand.

Linnaeus in Amsterdam

Two and a half years later, in 1735, Linnaeus travelled to Amsterdam, where he would meet some of the many influential people important to his career. His prospective father-in-law had insisted that he obtain his medical doctorate – something not available in Sweden at the time. This was little more than a formality, which involved presenting an already prepared disser-tation on fevers, answering a few questions and shaking a few hands. It took three days. It cannot be said that he was truly qualified to practise anything that we would call medicine, but in those days, who was? Always the self-publicist, Linnaeus heralded his arrival with notices in newspapers:

> *All that this skilful man thinks and writes is methodical*
> *and he does not rest until he has brought the science of the*
> *project, which he engages, into an order corresponding*
> *to nature. As can be concluded from this, he possesses*
> *an exquisite power of judgement, while not lacking a*
> *natural ability for invention. His diligence, patience*
> *and industriousness are extraordinary . . .*[87]

It seemed to do the trick, and in the space of just three years he went from being an unknown from the sticks to a central figure in the European scientific establishment. Hopeful of a publisher, he had taken with him to Amsterdam the manuscript of what was to become the first edition of *Systema Naturae*, in which he classified the natural world of animal, plant and mineral according to kingdom, class, order, genus and species.

One of his new associates, Johan Gronovius, was so impressed by *Systema Naturae* that he and another admirer, the Scottish doctor Isaac Lawson, paid to have it published.* The first edition was only fourteen pages long, and ran to a mere twenty-nine copies, two of which reside at the Linnean Society in London. It

---

* Isaac Lawson is commemorated in the genus *Lawsonia* (henna) and Gronovius in the flowering plant genus *Gronovia*.

was to continue for another fifteen editions, and would become one of the most important works on biology ever published: only Darwin's *On the Origin of Species* and Watson and Crick's 'Molecular Structure of Nucleic Acids' have had greater influence on the biological sciences.

Just to the west of Amsterdam was once the glorious garden of Hartekamp, the pride of an immensely wealthy banker, George Clifford. Clifford owed much of his wealth to colonial enterprises, which also afforded him the opportunity of importing a vast collection of exotic plants and animals. Linnaeus, often described as the 'second Adam', must have felt himself returned to Paradise when he first set foot in Hartekamp later in 1735. Twelve hundred plants from all over the world were recorded there, many of them in hothouses, providing Linnaeus with his first serious encounter with live exotic plants rather than dried specimens. There was also a zoo, which contained tigers, apes, peccaries and more. Clifford took Linnaeus in, ostensibly as his resident physician (although his only ailment appeared to be hypochondriasis), but chiefly to catalogue his botanic garden.

Linnaeus was given the run of the place, including access to the extensive library, which he was permitted to add to at will. He was also allowed to collect more plants for the garden. He flourished there, enjoying the benefit of a salary, board and lodgings plus time to write and time to study. The two major fruits of his two years with Clifford are *Musa Cliffortiana* (1736) and *Hortus Cliffortianus* (1737). The latter is a catalogue of the plants found at Hartekamp, while the former is a charming oddity – *Musa Cliffortiana* means 'Clifford's Banana'. This slim, large format book was a high-quality production destined to be an impressive gift to friends. Only a handful were printed. One might think that this was a vanity publication of show without substance, but it was from the pen of Linnaeus and is therefore an entertaining and enlightening read. In the hothouses of Hartekamp resided several banana plants, which had resolutely refused to fruit. Linnaeus, working on the sensible principle that a gardener should try to reproduce the native conditions of any plant to make it thrive, set the plant in rich soil (with the help of Clifford's gardener) and stressed it into flower by first starving it of water then watering it heavily while maintaining a tropically high temperature. Having induced the thing to flower, Linnaeus fretted over its sexual apparatus, noting that the female parts and male parts were never fertile on the same day, thus requiring more than one plant to effect fertilisation. Employing the over-romantic anthropomorphisms that made him a pariah among the strait-laced, he described it thus:

> *Therefore a peculiar and unexampled sort of polygamy occurs in this plant; in it, it is allowed that two different marriages may combine together; and the one wife, married to useless husbands, embraces the husband of the other female, while these husbands are united with a sterile incapable wife.*[88]

If Linnaeus was intrigued by this arrangement, he was dumbfounded by the plants' eventual lack of seeds, which existed only as shadows in the familiar black dots in the centre of the fruit. Linnaeus was not to know that the plant before him was a sterile hybrid of two wild species and that all that husband-and-wife nonsense was futile.

Aside from his poetic flourishes, *Musa Cliffortiana* is a comprehensive and detailed account of the banana plant. It also demonstrates much of Linnaeus's 'system' in action: he placed the plant within a hierarchy based on the sexual system that he had outlined in *Systema Naturae*; and he sketched what he considered to be a natural arrangement of plant genera. Incidentally, Linnaeus was content with the one-word name *Musa* for the banana because he had yet to establish two-word names for organisms (and with only one species in the genus it required no differentia in any case). It was the custom, soon to be a rule (of Linnaeus, who loved rules), that names (of genera at least) must derive from Latin or Greek or be eponyms for naturalists. Linnaeus was well aware that *Musa* came from the Arabic *moaz* for 'banana' and was therefore 'barbaric' (i.e. non-classical). But he liked the name and it did have some 'official' precedence in the works of Cesalpino and others. Cleverly he circumvented this troublesome etymology by 'naming' the banana after Antonius Musa, a botanist and physician to the great Augustus, thus coining an eponym that was also decidedly Latin. In time Linnaeus would provide a proper binomial for Clifford's banana: *Musa paradisiaca*, meaning 'banana from Paradise', reflecting the belief that it was the 'Tree of the knowledge of good and evil', and also named another species – *Musa sapientum*, the 'wise men's banana' – presumably on the same basis.

The banana has been a taxonomic nightmare for generations and Linnaeus could hardly have chosen a more problematic plant if he had tried. The specimens he examined were all cultivars and very likely hybrids of two natural species, now called *Musa acuminata* and *M. balbisiana*. *M. paradisiaca* lives on, as the hybrid it is, in *M.* × *paradisiaca* in accordance with the custom for naming hybrids (see p. 75). As a final twist, the original *Musa* from Clifford's garden was almost certainly a plantain and not what we would call a banana at all.[89]

In 1736, Clifford gave Linnaeus leave to visit England. The herbaria and botanic gardens of that fine country had been described to him by his friend Artedi, so Linnaeus was keen to see them and hoped also to be able to bring back live specimens for Clifford. *Systema Naturae* had not long been published, so the month-long trip would offer a further opportunity of promoting the work that he no doubt hoped would revolutionise taxonomy. As was usual with Linnaeus, he generally received an initial poor reception followed by overwhelming acceptance and praise.

The enormously wealthy Sir Hans Sloane is most commonly remembered for Sloane Square and for forming the collection that would be the basis of the British Museum and part of the Natural History Museum (although his introduction of drinking chocolate is enough to make him immortal in my eyes). Sloane's first meeting with the young Linnaeus was reported to have been cooler than it might, perhaps owing to his letter of introduction from the notable Dutch botanist Boerhaave. This suggested that Sloane – rich, learned, Baronet and President of the Royal Society – and Linnaeus – poor and unknown – were equals. However, Sloane showed Linnaeus around his collection, corresponded with him later and was memorialised by the young upstart in a genus of tropical and subtropical trees and shrubs, *Sloanea*, so relations obviously became cordial. The British Museum, established in 1753 immediately after Sloane's death, was among the first establishments to employ the Linnaean system.

In Oxford, at the Botanic Garden, Linnaeus met a much frostier reception from the Professor of Botany, Johann Dillenius. On his arrival at the garden he overheard Dillenius say of him: 'This is the man who had thrown all botany into confusion'. It is no surprise to learn that there is a genus *Dillenia*: the two became firm friends once Dillenius was convinced of Linnaeus's qualities as a botanist.

By 1736, *Systema Naturae* had already been published to a generally accepting public, together with *Fundamenta Botanica*, a companion work that laid out explicitly his sexual system – by which he categorised plants largely according to the numbers of stamens and pistils in their flowers, an idea that would gain him fame and notoriety in equal measure (see p. 202). But Linnaeus had only just started. In 1737, in addition to *Hortus Cliffortianus*, he published *Flora Lapponica*, an account of his earlier journey and discoveries in Lapland. That year also saw the publication of *Genera Plantarum* – a list of all the genera of plants arranged within the sexual system under classes and families – and *Critica Botanica* – an extension and elaboration of the earlier *Fundamenta Botanica*. These two works were later revised and combined to form his monumental *Philosophia Botanica*.

In 1738, when he heard the disquieting news that his betrothed, Sarah Elizabeth Moraea, was being courted by someone who had the advantage of actually being there, Linnaeus determined to make haste home to claim his prize before she succumbed to his rival's entreaties. Before he could leave he was taken ill from eating oysters, but then delayed his departure further by going home the long way via Paris to meet the de

Jussieu brothers* who were carriers of the Tournefort torch. Linnaeus arrived back in time and married his betrothed in June 1738.†

He may have been an established presence on the European botanical scene, but Linnaeus was now an unemployed presence, forced to seek work in Stockholm as a humble doctor of medicine. His practice struggled at first, until he met up with some of the well-to-do wastrels in the not-so-well-to-do parts of the city who were suffering from what the English call the French disease (and no doubt the French call the English disease). What they called it in Stockholm I do not know, but money was available to anyone able to cure it. Linnaeus employed the time-honoured method of mercury ointment (as his modern biographer Wilfrid Blunt put it: 'A night with Venus, a lifetime with mercury'). With no shortage of clients and a reputation for curing this and other debilitating diseases, he was soon in great demand. His medical career was, however, short-lived, and he soon found better things to do, returning in 1741 to Uppsala University, this time as a professor of medicine and later botany. Here Linnaeus was to remain until his death in 1778.

During the intervening thirty-seven years, he produced many more books, three in particular of significant interest: *Philosophia Botanica* of 1751, *Species Plantarum* of 1753 and the tenth edition of *Systema Naturae* of 1758, all of which had precursors. The first of these does what it indicates on the cover, laying down a practical philosophy for the naming and classification of plants, in

---

* Antoine Laurent de Jussieu (see p. 227) was their nephew.

† He might have fared better to let his rival have her. As revealed by the personal observations of Johan Christian Fabricius, a later Danish protégé and enormous admirer of Linnaeus, Sarah was large, domineering, selfish and without culture. The 'monandrian lily' (as Linnaeus referred to her), it seems, soon wilted.

particular the suggestion that two-word names for species would in all probability be a good idea. *Systema Naturae* and *Species Plantarum* mark the starting point for zoological and botanical nomenclature respectively, and for the first time use binomials consistently. It is in these three books more than any others that the Linnaean system is to be found.

The 'Linnaean system'

The 'Linnaean system' is still what people think of on the rare occasions they think about taxonomy at all. Certainly most people associate the name of Linnaeus with the familiar if unfriendly two-part Latin names of animals and plants. In his own time, through the sheer volume of his work and the unrelentingly pedantic nature of his rules, Linnaeus erected a comprehensive structure of classification and nomenclature. It was by no means perfect – and much has been eroded by modern thought – but it was good enough for the moment and, more importantly, became accepted by nearly everyone. At last, one naturalist could write to another about a species without ambiguity, and new species that were flooding into Europe from newly discovered territories could be readily described, named and allocated their spot in the great scheme of things.

Apart from dark references to a 'sexual system', I have given little away so far about Linnaeus's methods of classification and naming plants. His naming system comes down to us fairly intact, but his methods of classification bear only a superficial resemblance to the 'Linnaean system' as it is understood today.

All scientists are men or women of their time, and we might admire their work in some areas while other aspects of their science or world view may seem absurd to the modern eye. Linnaeus had ideas that seem ridiculous to modern science, although they were universal at the time. As some of the more scripture-literal Christian commentators never tire of pointing out, he was a creationist – which he was, although he would have considered himself just a Christian. Darwinian evolution was a century in the future and what little evolutionary thought there had been up to Linnaeus's time had made almost no impact. So it was without the slightest fear of contradiction that he wrote in *Philosophia Botanica*: 'We maintain that, in the beginning of things, a single sexual pair of every species of living being was created.'[90]

Constrained in his thinking, like his predecessors, by these notions, Linnaeus believed that any groupings of species within genera and groupings of genera within families and so on were the result of a divine plan. He considered it his task to discover what that plan was: '*Deus creavit, Linnaeus disposuit*' ('God created, Linnaeus ordered'), he was fond of saying.*

The Russian evolutionary biologist Theodosius Dobzhansky famously said: 'Nothing in biology makes sense except in the light of evolution.' In as much as he meant Darwinian evolution in the broad sense (it is a theory made up of several theories rolled into one) I think he was, if not wrong, then not quite right, either. It would be more accurate to say: 'Nothing in biology makes sense except in the light of common descent.' The notion of common descent was a long time coming considering how blindingly obvious it is in retrospect. Fossils were already well known in the eighteenth century – Linnaeus even had a class

---

* It is worth noting that he did come to accept that species could change by hybridisation, although he considered genera to be fixed.

Fossilia within his classification of the 'kingdom of minerals' – but their existence was believed to represent previous creation events, not the ancestors or (much more often) the relatives of ancestors of modern species we now know them to be. The very existence of genera and higher taxa would, one might expect, suggest the simple explanation that they had similarities because they were related by common descent. But almost no one made this deduction. Linnaeus certainly did not, and it is impossible to make a classification in the absence of an accurate theory on which it may be based. If one does not know the reason why things appear to be ordered in some way, one is likely to get the ordering wrong. That his system of classification came as close as it did to the real arrangement of the natural world is due to the fact that things related by common descent tend to look similar. Part of the reason that he got some things wrong is because sometimes they don't.

While creationism is now beyond the pale in polite scientific circles, philosophy (at least some philosophy) finds a welcome home in science (at least some science). Philosophy is essential for mathematicians, an aid to physicists, worthless to chemists and a disaster for biologists. Linnaeus, although brought up on Platonic and Aristotelian logic, was a pragmatist and did not fall into the trap of attempting to apply a strict logical scheme to a real world that would, he knew, have none of it. Aristotle (see p. 145) himself had difficulty applying a Platonic structure to his own taxonomy – the world of biology, he found, is far too unruly to be constrained within any philosophy; Linnaeus would have faced the same problem. It is often suggested that Linnaeus did apply Platonic ideas of essentialism – his writings are certainly full of terms such as 'differentia' and 'essence' – but this idea has met with opposition among some contemporary commentators. All I can add to the exchange between academics on this subject is that if Linnaeus did try to apply essentialist logic to taxonomy, he was not terribly good at it.

Natural versus artificial systems of classification

As a novice botanist, my first book was a lucky choice. It is called *Wild Flowers of Britain and Northern Europe* by Fitter and Blamey. Despite having acquired another thirty such guide books, it is always to Fitter and Blamey that I first refer when attempting to identify a plant. It has something that the others lack: a simple, visual key.

At the front of the book are ten pages devoted to pictures of flowers. They are arranged in a hierarchy. The first category is for open flowers (such as buttercups, primroses, etc.). This is divided up into five further subgroups: one petal, two petals, three petals, four petals, five petals and six or more petals. Each of these, in turn, is divided into pictures showing what colours and shapes are available in that subgroup. If I want to identify an open flower with five broad, blood-red petals, a quick scan of the page will show that it is either hound's-tongue or hawkweed saxifrage.

This is a taxonomy of the flowering plants of Britain and Northern Europe. But it is not a *natural* taxonomy. A natural taxonomy would not do what my book does. It would not, for example, place navelwort, bilberry, deadly nightshade and solomon's seal together. These four plants are very different and belong in different families. The system of classification in my book is artificial. But it works, which is all that anyone can ask of it.

Linnaeus was a teacher, and much of what he wrote was teaching material. One of his primary motives, therefore, was to enable his students and others to rapidly identify plants. This he did in a way not dissimilar to Fitter and Blamey – he based his system on a hierarchy of floral characters. But, of course, it was not a natural system; it was self-consciously created as an artificial one.

The sexual system

All taxonomic classifications are complicated, and those of Linnaeus are particularly eye-watering. He carefully differentiated between artificial systems, mere matters of convenience, and a natural system, which reflected the actual arrangement of living things. His most famous taxonomy of plants – the sexual system – was self-consciously artificial, often lumping together totally unrelated genera and separating those that were closely related. It was also notorious.

In *Systema Naturae* and elsewhere, he divided the natural world into three kingdoms: minerals, animals and plants.* These kingdoms he further subdivided into class, order, genus and species, genus being by far the most important taxon for Linnaeus, with everything else revolving around it. His way of classifying plants and animals within these taxonomic rankings was different, so I will tackle the plants first.

Linnaeus looked to the reproductive parts of plants to provide him with what he called 'factitious characters' – distinct morphological characteristics that could be used to form groups. His twenty-four 'classes' were defined chiefly by the male parts of flowers – the tiny, lollipop-shaped, pollen-covered structures called stamens. Monandria (literally 'one male') is the class of all plants with one stamen; Diandria, all those with two stamens and

---

* This book is concerned only with his views on animals and plants.

so on, up to the Dodecandria (with twelve to nineteen stamens) and Polyandria (many stamens). The other classes are based on further personal arrangements within the flower, such as Polygamia (both hermaphrodite and unisexual flowers on the same plant) and Dioecia (male and female organs on separate plants).

His 'orders' within those classes defined by their numbers of stamens were differentiated by the number of female parts (pistils). The number of these is reflected in the names he gave them: Monogynia, Digynia, Trigynia through to Dodecagynia and Polygynia, with a few numbers not represented – there was no Octogynia, for example. Within the other classes, Linnaeus relied upon a ragbag of characteristics to define orders. The class Didynamia, for example, was divided into two orders still familiar today: Gymnospermae and Angiospermae, having four naked seeds and four enclosed seeds respectively.

There is considerably more to it than this, but I will trouble you no further. Except to consider his twenty-fourth class, Cryptogamia,[*] which he divided into four orders: Filices (ferns), Musci (mosses), Algae and Fungi. Ferns and mosses are certainly plants, fungi are categorically not, and algae have never been able to make up their minds what they are. Few people would blame Linnaeus for giving up on the lot of them, but mycologists have never entirely forgiven his abandonment of the fungi. He described only about eighty species in *Species Plantarum* – about the same number of fungi that I find on a good-to-average fungus foray. It is not as though nothing was known about fungi in

---

* *Cryptogama* means 'hidden marriage', reflecting the fact that, unlike the brazen flowering plants, their means of reproduction is kept coyly from sight.

Linnaeus's time. Pier Antonio Micheli, in his *Nova plantarum genera* of 1729, had listed nine hundred species (some of which were, admittedly, lichens),* illustrating many with sufficient accuracy to make them instantly identifiable. He also studied fungi using a microscope and wrote about their spores and mycelial growth. He even showed how fungi can grow from spores to produce new fruiting bodies (mushrooms).

One might assume that Linnaeus never saw Micheli's book, but in *Philosophia Botanica* he wrote: 'The class of the Funguses has been arranged by Dillenius and Micheli.' On one of my visits to the Linnaean Library, I asked to see a copy of the Micheli. They had only one – Linnaeus's own – and it was in remarkably good condition, almost as though it had hardly been opened. Pleased with this idea, I suggested it to the librarian, but she told me that it had simply been extensively conserved. Maybe Linnaeus didn't much like what he read. Micheli, incidentally, is often called the 'father of mycology'; quite right too.

Selling the system

For the plants, if not the fungi, Linnaeus's sexual system was popular because it provided a reliable and, believe it or not, easy way of identifying plants. One might wonder how he managed to sell what seems to be a method of baroque and unmanageable complexity; I suggest it is partly because all it requires a botanist to do is to count things in order to identify a species, and partly down to Linnaeus's salesmanship. As all advertising agencies know, nothing sells like sex, and Linnaeus sold it like no one

---

* Lichens are composite organisms consisting of a fungus and a photosynthetic alga or cyanobacterium. Although they are informally referred to with the name of the fungus, in truth they have no names.

before him. He called it the 'sexual system' and that is what, in greater part, it was. But calling it such and making the odd reference to 'male stamens' or 'female pistils' would not have delivered the required amount of titillation. Linnaeus instead dressed his botanical descriptions in outrageous erotic matrimonial metaphor. The fact that they were kept within the marriage bed makes them more shocking, not less.

The slightly intimidating technical words that Linnaeus used for his classes and orders are merely Latinised Greek words for rather homely concepts. In using the Greek words *andria* and *gynia* ('man' and 'woman'), he had settled on the idea of denoting stamens as husbands and pistils as wives (or sometimes concubines). *Genesis* is, of course, 'birth', *delphus* 'brother' and *oecus* 'a house' or 'home'. With this small vocabulary, enhanced by neologisms constructed with the addition of qualifiers (*Diandria* meaning 'two men', for example), Linnaeus was able to enjoy himself with a large number of unconventional domestic arrangements about which he was quite specific.

As I mentioned earlier he referred to his own wife as his Monandrian lily, reflecting the fact that she had only one husband. A slightly racier example is the order Digynia of the class Monandria, representing one husband with two wives. The order Monogynia of the class Polyandria is 'one wife with many husbands'; the order Monadelphia of the class Dioecia is a tricky one to translate, but I am going with 'close brothers living together in one house, married (en masse) to any number of wives living in another house'.

It is great fun playing with these names: how about order Polygamia Frustranea of the class Syngenesia? This could be translated as 'husbands with their private parts fused together belonging to several [understandably] frustrating marriages'. If you think I am reading more into this than Linnaeus intended, here is one of his own: Enneandria Monogynia. He described it as 'six men in the same bride's chamber, with one woman'.[91]

Linnaeus was quite clear about his metaphors. His evocative names of classes and orders were his own and in *Philosophia Botanica* his intent is laid bare:

> *Therefore the Calyx is the bedroom, the Corolla is the curtain, the Filaments are the spermatic vessels, the Anthers are the testicles, the Pollen is the sperm, the Stigma is the vulva, the Style is the vagina . . .*[92]

This is bad enough, but it gets worse. He wrote that the calyx 'could also be regarded as the lips of the *cunnis*'. Faintheartedness forces me to revert to the original Latin for the final word, but Stephen Freer, who provided the English translation, was fearless.

The eighteenth century saw the publication of *Fanny Hill* and a thousand other erotic titles, but few expected to find pornography within the pages of a book on plants. Pornography was for drinking clubs, not the Royal Society. While Linnaeus's system was accepted by many naturalists as a comprehensive guide with amusing sexual allusions, others were horrified, even refusing to believe that plants could indulge in sexual activity at all, let alone such wanton behaviour as Linnaeus had so colourfully described. Linnaeus was also criticised for some of his fruitier plant and animal names. His publications were banned from the Vatican by the Pope. But the most vociferous objections came from a fellow naturalist, Johann Siegesbeck, a German academic from St

Petersburg, who described Linnaeus's work as 'loathsome harlotry as several males with one female would not be permitted in the vegetable kingdom by the Creator!'

Siegesbeck ridiculed the sexual system while Linnaeus was in Holland, and this may have caused Linnaeus difficulty in finding an academic position when he returned to Sweden. Whether or not there were practical repercussions, Linnaeus resented the criticisms and never forgave Siegesbeck, complaining that 'it is an idiot and fool, under which I have suffered'. He duly named *Sigesbeckia* – a genus of 'small-flowered weeds', as described by the modern taxonomist William Stearn – and one day, on finding a packet of *Sigesbeckia* seeds, he wrote on the packet the name *Cuculus ingratus* ('ungrateful cuckoo'). There was at that time much interchange of plant material between Uppsala and St Petersburg, and unfortunately (or not) the packet fell into Siegesbeck's hands.

The most extensive criticism of the sexual system came from the first edition of the *Encyclopaedia Britannica* in the late eighteenth century, written by one William Smellie. He did not like Linnaeus's ideas at all and devoted more than ten thousand words to explaining why. The first nine and a half thousand words consisted of a well written and thoughtful critique of the very existence of sex in plants – all of it wrong. Part of his argument was that sex in plants is impossible because plants are not alive. Indications of life, such as the common observation that flowers in a window often follow the sun, he dismissed as fanciful, suggesting that the phenomenon was due to directional wilting. The last five hundred words directed unmitigated fury at Linnaeus for his sexual metaphors, suspecting even his motives. 'One would be tempted to think that the author had more reasons than one for relishing this analogy so highly,"* he wrote. 'In many parts of this treatise, there is such a degree of indelicacy in the expression as cannot be exceeded by the most obscene romance-writer.' He continued: 'Men or philosophers can smile at the nonsense and absurdity of such obscene gibberish; but it is easy to guess what effects it may have upon the young and thoughtless.'

Among philosophers and the thoughtful at least, Linnaeus's sexual system was quite a hit in its time, putting plants in some sort of order. Before Linnaeus, names were absurdly long, and there was no accepted system of classifying them. Biology was like a shop in which things had been placed on shelves in almost random order, often with the wrong label and nearly always with more than one. His system was artificial – and therefore 'wrong' – but, like my favourite wild flower guide, it produced results quickly and reliably. A big boost to its fortunes came when William Hudson used it systematically in his *Flora Anglica* of 1762, and it dominated botanical taxonomy until the early nineteenth century.

Despite its general popularity and usefulness, it was obvious that the sexual system was a poor substitute for a natural system because it tended to split or lump natural groupings randomly. Species belonging to a single genus cannot always be relied upon to produce the same number of stamens, for example: the genus *Verbena*, which Linnaeus placed in the class Diandria because its species usually have two stamens, contains many species with four stamens. However Linnaeus allowed his sense of affinity, his sense of a natural order, to triumph over system and put the lot into Diandria regardless. He is quite notorious for tweaking his

---

* In this, at least, one must suspect that Smellie was correct.

classifications. Where he has not done so, double-entries (the same genus inhabiting different orders, for example) occur.

Throughout his working life, Linnaeus attempted to construct a truly natural system of his own. 'Fragments', as he called them, of his natural system appeared in various publications over the years, including *Philosophia Botanica* in 1751: 'The fragments of the natural method are to be sought out studiously. This is the beginning and the end of what is needed in botany'.[93] He used the word 'fragments' with good reason, for this was not an outline of a hierarchy, just a list of sixty-eight orders with their associated genera. *In Philosophia Botanica*, at last, his orders have appropriate names, replacing the treacherous artificial names of the sexual system. Several, such as Caryophyllei (Caryophyllaceae, or pinks) and Cucurbitaceae (gourds), are familiar as modern families. The order names are mostly descriptive, but not in the way they were in the sexual system; they say something more usefully characteristic of the order. For example: Drupaceae (including plums) bear 'drupes'; Luridae (mostly tomatoes) often have pale yellow flowers; Compositi (daisies and allies) have composite flower heads and so on.

Linnaeus apologised for his 'fragments', writing in *Philosophia Botanica*: 'The absence of things not yet discovered has acted as a cause of the deficiencies of the natural method; but the acquisition of knowledge of more things will make it perfect.'[94] Elsewhere he declared that discovering the natural order would be a thing for future naturalists, although he doubted it would ever be accomplished. I wonder what his opinion would be of the monster that is modern classification, or of the fact that there is a true natural order that goes beyond similarity of form.

My friend Bryan would be the most annoying person I know, were it not for the fact that he is gentle, charming and a joy to be with. His insufferable quality is that his has casually committed to memory the common name, Latin name, synonyms, description, habitat and history of very nearly every living thing in England. Carefully examining a liverwort on a tree stump with a lens, he will tell you that it is *Cephalozia lunulifolia*, mentioning in passing that the twittering from a nearby gorse bush, *Ulex europaeus*, is that of the cirl bunting, *Emberiza cirlus*, adding that it is a rare species in Dorset and more common in south Devon. He is the ultimate generalist, proficient in all of natural history. But if you ask him what he calls himself, he might say that he is a lichenologist.

Linnaeus, too, was a great generalist, but if anyone had asked him what he what he called himself, he would have said that he was a botanist. In his extraordinary self-imposed mission to classify everything that lived, he had to tackle his lesser 'speciality', the animal kingdom. It is an easy amusement to mock his resulting treatment of the animals (a temptation I will not entirely resist), but it is important to remember that he was working within the severe limitations of his time. Common descent was virtually unsuspected, the *scala naturae* was still a powerful paradigm and comparative anatomy – the study of the morphological differences and similarities between animals – was in its infancy. Linnaeus did not have the knowledge or indeed the time to come to anything like the conclusions we have arrived at today. He also fell into the common and disastrous trap of choosing the wrong defining characteristics for his higher taxa.

Linnaeus arranged the animal kingdom using the same four categories that he used for the plants: class, order, genus and species. He devised six classes: Mammalia (we have Linnaeus to thank for this word), Aves (birds), Amphibia (frogs, reptiles, lampreys, sharks), Pisces (fish), Insecta (insects, spiders and crustaceans), Vermes (molluscs, worms, coral, some seaweeds and just about everything not covered in the previous five classes). He arrived at these six by looking first at the structure of the heart, which divided the classes into three unnamed groups: Mammalia/Aves; Amphibia/Pisces; Insecta/Vermes. He split Mammalia from Aves by them being viviparous or oviparous (live-bearing or egg-bearing); Amphibia from Pisces by their mode of respiration (lungs or gills); Insecta from Vermes by their possession of antennae or tentacles. These classes were each divided into between three and eight orders, such as those exemplified above by the animals in brackets.

Even at this high taxonomic level it is clear that Linnaeus's classification bears almost no resemblance to what we now know to be the true order of the animal kingdom. For example, four out of six of Linnaeus's classes were vertebrates, whereas the Chordates (as the modern and much enlarged group is now called) is one of more than two dozen phyla. Vermes is what is now called a 'waste-basket taxon', accommodating everything that does not fit anywhere else. It is one hell of a waste basket, however, containing the vast majority of known phyla. It is particularly disconcerting to find some seaweeds in this class, because we now know these to be algae and not animals. (The seaweeds in question are the chalky Coralline species, which form structures not unlike sea fans, hence the confusion.) However, it is at least gratifying to see that Linnaeus placed whales where they belong, with the other mammals.* Earlier he had considered them

---

* Wilfrid Blunt reports that on the wall outside Linnaeus's bedroom hung a picture of a whale with its offspring still attached by an umbilical cord, irrefutably placing the whales in Mammalia.

to be fish – an allocation rather forced upon him by his previous practice of calling Mammalia the 'Quadrupedia'. No doubt he had thought that anyone classifying an animal among the 'four-legged' while it clearly had none would not have a leg to stand on.

The naming of man

Also in Mammalia is a species that Linnaeus called *Homo sapiens*. Considering some of the names given by thoughtless taxonomists to their species, we have done rather well and perhaps better than we deserve. *Homo sapiens* means, simply, 'wise man' and is likely to be a reference to Aristotle's concept of the rational man. *Homo* is of course Latin for 'man', ultimately, it is believed, derived from a Proto-Indo-European word reconstructed by imaginative paleo-linguists as *dhghem*, meaning 'earth'. The association of man and earth goes back to Adam himself, a sobering reminder of both our origin and our destination. *Sapiens* is more straightforward, being from the Latin *sapere*: 'to be wise'.

There are around two hundred members of the mushroom genus *Agaricus*, and a different dandelion (*Taraxacum*) for every day of the year, but we are quite alone in the world as the only extant member of *Homo*. There have been suggestions that our nearest living relatives, *Pan troglodytes* and *P. paniscus*,[95] the chimpanzees and indeed other apes, should be assigned to *Homo*, but this is unlikely to happen. Having once seen a mother chimpanzee tickling her baby with all the gentle teasing and love shown by human mothers, I have some sympathy with this view, but there are at least twelve million years of divergence between us (counting down and back up again) filled with many inter-mediate species. The distance is too great.

There have, however, been several other *Homo* species over the ages, and it has been the life's work of many palaeontologists to discover and describe them. Not that they often agree with one another; the whole area of recent human evolution is filled with (reasonably healthy) controversy and it is not easy for the layman to get a straight answer to even the simplest of questions. Nevertheless a vague consensus (which may be swept away at any time) has formed.

The generally accepted view is that the hominids (*Homo*) originated in Africa and that some of them spread outwards to Asia and Europe in several waves with varying degrees of success. The first *Homo* was *Homo habilis* (delightfully, 'handy man'), followed by *Homo ergaster*, 'working man'. The line then divided, producing the ultimately doomed *Homo erectus*, 'upright man', on the one hand and our own line on the other. *Homo erectus*, largely found in Asia, appears to have itself divided, producing the recently discovered 'hobbit', *Homo floresiensis*, 'man from Flores'. This latter is likely to be our last surviving relative, the remains being only twelve thousand years old.*

Following *Homo ergaster* in our own part of the family tree was *Homo heidelbergensis* (named, with no consideration for collectors of interesting appellations, for the University of Heidelberg), after which the line splits again to give *Homo neanderthalensis*, 'man from the Neander Valley', and us. I have always felt sorry for Neanderthal Man, who was, after all, just a well-built, larger-brained version of us, complete with art, burial of the dead and clothing. He faced death and ultimate extinction,

---

* There is, however, a tradition among the inhabitants of the island of Flores of a 'wild man', the Ebu Gogo, surviving until modern times. We may only just have missed them.

very possibly at our hands either by direct action or by usurpation of resources.

Nevertheless, we did manage to live side by side for tens of thousands of years, although with what degree of interaction is not really known. In the Middle East, evidence exists of both species being present in the same locality over the same period. However, we are warm-adapted and they were cold-adapted, so we may both have indulged in a little timeshare as the climate warmed and cooled. Until recently, the small amount of DNA testing that has been done suggested that the sometimes mooted romantic alliances between the two species never happened, or at least, never produced fertile issue. However, a heroic effort to map the Neanderthal genome indicates that it has occurred, though not very often and possibly only once, at what one palae-ontologist calls a 'love-in'.* (Having seen an artist's impression of what a lady Neanderthal looked like, I can't say I am surprised that such liaisons were rare.) This small amount of distinctively Neanderthal DNA appears to have found its way into the genetic makeup of only those *Homo sapiens* that are not of sub-Saharan populations, so only some of us have the cold-loving Neanderthals as ancestors. Their last-known bastion was in southern Spain at least 28,000 years ago, after which they disappeared from history.

There are several other *Homo* found in the literature, but for the purposes of this book the most important is *Homo sapiens idaltu*, 'wise elder', a subspecies that lived in Ethiopia around 160,000 years ago. It has also been strongly argued that

---

* It has been suggested that it happened a great deal more often than genetic analysis might indicate, as resulting lineages may have been selected out due to a compromised genome.

Neanderthals, too, should be a subspecies: *Homo sapiens neanderthalensis*. In either case, we automatically become a subspecies and, under the rules of nomenclature, must endure the mild taxonomic tautology of *Homo sapiens sapiens*. This appellation certainly does not indicate that we consider ourselves to be doubly wise; it is just a matter of taxonomic convention.

Linnaeus's take on humankind and its place in the natural order was an uncomfortable mix of good sense and bad. Bravely – Darwin's *On the Origin of Species* was still a whole hundred years away – Linnaeus made a break with his taxonomic predecessor, John Ray, and placed us squarely with the animals on his consistently used basis of shared characteristics. His nod in the direction of the angels was to give us the specific epithet *sapiens*. Comparative anatomists of the seventeenth and eighteenth centuries, such as Pierre Belon and Richard Bradley, who had studied the anatomies of apes and men, had left Linnaeus little serious choice, although he showed no sign of wanting one: 'And the fact is that as a natural historian I have yet to find any characteristics which enable man to be distinguished on scientific principles from the ape'.[96]

Nevertheless he did suffer criticism for including us with mere animals. Johan Gottschalk Wallerius, a rival of Linnaeus's for a medical professorship, noted that there were a large number of differences between man and apes, or, more particularly, mammals. He demanded reasons for why cattle with four legs, monkeys and apes with four arms, manatees with two arms/no legs and whales with no limbs at all should all be lumped together in Linnaeus's *Quadrupedia*, which, after all, means four-footed. These arguments seem foolish now, but the idea that manatees or whales once possessed a full complement of limbs only makes sense after Darwin; it had crossed the mind of no one in the mid-eighteenth century. There were also rather absurd complaints about Linnaeus's class Anthropomorpha ('shaped like a man'), a group that contained sloths, monkeys and apes, including man. But how can man be like himself, it was argued. As the transparently tendentious nature of the arguments suggests, the main concern of the critics was clearly the dignity of man and, more important still, his spirituality – concerns that were to be revisited with renewed force a century later.

After Linnaeus's sensible start things went seriously awry. His treatment of *Homo* seemed to be based more on myth and legend than fact. It changed confusingly with every edition of *Systema Naturae*, and introduced obscure synonyms. His genus *Homo* was broader than that used today, because it included all the apes, not just humans. He recognised several species of *Homo* and also two overarching groups akin to subgenera: *Homo diurnus* and *Homo troglodytes* ('man of the day' and 'man of the caves'). The latter was given the parallel name *Homo nocturnus* – 'man of the night' – followed by several other 'helpful' synonyms, such as *Homo sylvestris* and *Orang Outang*. The latter two names, both meaning 'man of the woods', would seem to solve the puzzle of what exactly *H. nocturnus* was, but a glance at Linnaeus's description indicates otherwise. He wrote that its body and hair were white, not orange, and that it had a *membrana nictitante* – membrane over its eyes, like that of an owl – which the orang-utan we know today does not. Nor, as far as we can tell, does this fine animal 'believe that the world was made for him, and would eventually be restored again to his sole dominion'.* The chimpanzee, which inherited Linnaeus's specific epithet and is now *Pan troglodytes*,† would be a contender, but obviously it does not fit the description either.

The *scala naturae* and its principle of plenitude was important to Linnaeus, so the opportunity to fill what he perceived as a gap between us and the apes had to be grasped. The white skin and hair give the clue to the nature of this odd creature. There had been many reports of such an animal over the centuries, not least in Pliny, and Linnaeus spoke to people who assured him that they had seen one, or at least spoken to someone else who had. Linnaeus expended much effort in trying to obtain a specimen, unsurprisingly without success. This creature was born of myth and traveller's tales wrapped around a kernel of truth. *Homo troglodytes* is simply an albino Negro and Linnaeus's description was lifted directly from, of all people, Voltaire, who had described one that had been displayed in Paris in 1744.[97]

The great apes were little understood in Linnaeus's time, and the chimpanzee and orang-utan found themselves sharing the name *Simia satyrus*.‡

Meanwhile, Linnaeus also described the mysterious *Homo caudatus* – 'tailed man'. In the twelfth edition of *Systema Naturae* (1766), he called it Lucifer and wrote it as a footnote. Tales of tailed men had appeared throughout history, and babies are sometimes born with tails (although nowhere near as often as even modern myths suggest). Human tails are caused by a variety

---

* '. . . *credit sui causa factam tellurem, se aliquando iterum fore imperantem*'. From the twelfth edition of *Systema Naturae* (1766).
† Although it is an uncomfortable name for a species that spends most of its time in the trees, as it means 'cave dweller'.
‡ An earlier name was *Satyrus indicus*, 'Satyr of India', although neither animal hails from those parts.

of medical conditions and are not throwbacks to earlier times. An entire 'species' of men sporting them has never been discovered, and the travellers' stories on which Linnaeus based them were merely garbled accounts of other primates. An extraordinary drawing appears in Linnaeus's *Anthropomorpha** of 1763 depicting a bandy-legged quartet designated as 'Troglodyte', 'Lucifer', 'Satyr' and 'Pygmee'.

The imaginary *Homo troglodytes* has a chequered history, but the real *Homo sapiens* fared much worse. Linnaeus was a man of his time and his discussion of this group would make even Joseph Goebbels blanch. He divided *Homo diurnis* into two broad species: *Homo sapiens* and *Homo monstrosus*. The latter was an extraordinary ragbag of sideshow exhibits. Among them were the 'races' Alpini, timid Alpine dwarves, and Patagonici, indolent Patagonian giants. The most puzzling – the Canadenses – were, apparently, flat-headed Canadians, and the most fantastical – the Monorchides – turn out to be Hottentots (gloriously, *Hottentotti* in Latin), who are '*ut minus fertiles*' ('less fertile'), owing to a deficiency of one in the trouser department. Better than Goebbels, at least.

Homo sapiens, too, has much to resent in Linnaeus's treatment. The precise allocation of species, subspecies and variety is confused and confusing, but overall there were five groupings. The strangest of these is *Homo sapiens ferus*, the 'wild man'. This was defined as being four-footed, mute and hairy. Several examples were given, such as the 'Lithuanian Bear Boy', the 'Wolf Boy from Hesse' and the 'Wild Girl of Champagne'. These are no more than unfortunate lost children, such as have existed

---

* The author was actually Christian Emanuel Hoppe, but Linnaeus's name is on the cover and it was written under his direction.

throughout history. The remaining four varieties or races were assigned, for reasons we can only guess at, the properties of the 'four humours' of Greek medicine. The 'choleric' Americanus was, among other things, beardless, obstinate and regulated by customs; the 'melancholic' Asiaticus severe, haughty and covetous and governed by opinions. The 'phlegmatic' Afer (African), unsurprisingly, did not come out of this well, being considered crafty, lazy, negligent and governed by caprice, while the 'sanguine' Europaeus was described as being gentle, acute, inventive and governed by laws. To his credit Linnaeus did indicate an understanding that differences between 'races' were a matter of culture and not nature, for he wrote that they vary 'by education and situation'.

The Linnaean approach to man is unthinkable now; all concepts of race are increasingly unfashionable and that of variety (in the technical sense) or, heaven forbid, subspecies, beyond contemplation. If we were flowers, however, botanists would not hesitate to split us into numerous varieties. With black Africans considered the standard, white Europeans could be *Homo sapiens* var. *alba* and so on.

But humans are animals and one must wonder if zoologists have short-changed us. Another animal that arose in Africa, the lion, has been divided up into several subspecies, despite them all looking remarkably similar to the non-specialist. There is certainly not the highly distinctive difference between subspecies of lion that we see between different populations of humans. Michael A. Woodley, in a remarkable paper published in 2009, noted that the genetic range across four subspecies of chimpanzee (*Pan troglodytes*) is very similar to that across the single species *Homo sapiens*, and argued that a consistent taxonomic treatment would provide us with five subspecies.[98] Should this treacherous, if scientifically legitimate, path be followed, then white Europeans would retain *Homo sapiens sapiens*, because the original description was based on a white European type (see below) – not because they are more important. The rest of mankind would acquire subspecific names, such as *Homo sapiens africanus* and *Homo sapiens orientalis*.

To be honest, I rather like the idea because it glories in our heterogeneity rather than pretending that it does not exist. However, while few areas of knowledge should be taboo, there are perhaps some things with which we cannot be trusted.*

Although not the custom in Linnaeus's day, every described species now requires a type (see p. 115) in order to tie the name to a physical object. For a variety of reasons, not all species have a type, and Linnaeus did not retain one for *Homo sapiens*. The story that the nineteenth-century American palaeontologist Edward Cope offered himself as the type specimen is simply untrue, despite its popularity. Ideally the sample should be the one most closely studied by the author. I have a fancy that Mrs Linnaeus might have been the best choice (although according to hints in his letters, Linnaeus might not have had the opportunity to study her as closely, or at least as often, as he wished). In 1959, as a passing remark in a paper on Linnaeus, the great taxonomist William Stearn wrote: 'Linnaeus himself, must stand as the type of his *Homo sapiens*'.

This counts as a nomenclatural act and under the rules is enough to make it so. Linnaeus did not designate himself, so he cannot be a holotype (see p. 119); however, he was one of the

---

* Maybe Linnaeus's aphoristic plea at the start of his treatment of *Homo* – *Nosce te ipsum* ('Man, know thyself') – is peculiarly apt here.

specimens that he examined, and as such was designated by Stearn as the lectotype. The type specimen of *Homo sapiens* now lies peacefully beneath Uppsala Cathedral.

Nomenclature

If Linnaeus had presented his ideas on taxonomy and nomenclature only in theoretical papers, little notice would have been taken of them. What truly brought his methods to life was the fact that he applied them himself on a monumental scale. Nowhere is this more evident than in the number of species he named. The biologists I have met who have named a species or two themselves are extremely proud of their creations, and rightly so. However, Linnaeus named some 4,400 species of animal and 7,700 plants, producing a body of work and a practical basis from which taxonomy could develop.

By consistently employing binomials in *Philosophia Botanica* in 1751, *Species Plantarum* in 1753 and the tenth edition of *Systema Naturae* in 1758, he established an irresistible precedent for their use. However, even in these and later works, Linnaeus did not consider binomial names to be true names. His 'complete' name for the foureye butterflyfish, for example, was: *Chaetodon spinis pinnae dorsalis 12, corpore striato, ocello subcaudali* – a phrase name. Meaning 'thing with flowing, tooth-shaped spines on its back looking a bit like a wing, a stripy body and a small eye just in front of the tail', it is a great description of the fish, making it instantly recognisable.

But it is still a trial to both memory and speech; where is the binomial of which we hear so much? It is not provided here or anywhere else in Linnaeus as a straight binomial, but instead in the following manner as one of the species of the genus *Chaetodon*:

capistratus 15. C. spinis pinnae dorsalis 12, corpore etc.

The number '15' means that it is the fifteenth species of *Chaetodon* described in the book. 'C.' is an abbreviation of *Chaetodon*. Appearing in the margin, 'capistratus' (stripy head) – now universally accepted as the specific epithet – is, for Linnaeus, just a quick way of referring to this member of the genus. Linnaeus, so fussy and so governed by rules, introduced his most famous, most useful 'rule' as a mere piece of shorthand. *Chaetodon capistratus* became its name and that is the name it has retained.

In *Philosophia Botanica*, he provided the rule that seems to define the binomial system: 'Every plant-name must consist of a generic name and a specific one.' But the specific name in this definition is not the single word, it is the phrase name. He did, however, suggest a single word for what we now call a specific epithet, the *nomen trivialis* or 'trivial name', an appellation sometimes applied today. Importantly, it was not used in the sense that it was trivial in comparison to the generic name, but that it was trivial in comparison to the full name. Then, as a mere footnote, he wrote: 'Trivial names can perhaps be allowed in the manner in which I have used them in *Pan Suecicus*;* these would consist of a single word'.

Linnaeus believed that a naturalist should remember as many names as possible, and in the binomial system he did more than anyone else to accomplish this goal. Not surprisingly the idea became extremely popular almost immediately, but if the binomial system was all that Linnaeus was remembered for he would be remembered little – a bit like the fellow who devised the brilliant stylised map of the London Underground, now

---

* A book published in 1749.

immensely useful in underground systems all over the world. Few can remember the name of the inventor,* although the fact that he received five pounds for his trouble does stick in the mind rather.

Linnaeus did not just invent the form of biological names; he invented the whole naming process. Most of his rules, and there are a lot of them, appear in *Philosophia Botanica* as a series of remarkable aphorisms. Together they form the basis of modern naming procedures.

Priority, one of the fundamental principles of taxonomy (see p. 130), was first enunciated by Linnaeus. Aphorism 243 says it all: 'A generic name that is satisfactory may not be changed for another, even if the latter is more appropriate.' In 257, he tentatively provided us with the freedom to name as we wish. Specific epithets may come 'from any source', he wrote, and (in 256) be 'free from any laws' – thus the specific epithet has none of the rules governing higher taxa.

Some of Linnaeus's aphorisms have been honoured more often in the breach. Aphorism 211, for example, indicates that names should be sensible: 'Only genuine botanists have the ability to apply names to plants,' he wrote, adding that 'private individuals have applied absurd names'. As amply demonstrated in Chapter II, genuine botanists (and zoologists, who are the worst offenders) are perfectly capable of inventing ridiculous names without the assistance of amateurs. Perhaps aphorism 249 is the most abused of all. 'Generic names one and a half feet long, those that are difficult to pronounce, or are disgusting, should be avoided.'

---

* Let us remember him now. He was Henry Charles Beck, known as Harry Beck, and he died in 1974.

Amphibalanus improvisus

CHAPTER VIII

TAXONOMY *after* LINNAEUS

L INNAEUS'S SEXUAL SYSTEM EVENTUALLY WENT OUT OF FASHION AND his animal taxonomy was superseded. But he had put things in order; not necessarily the *right* order, but the wrong order is sometimes better than nothing. Certainly it was better than the chaos that threatened to overwhelm taxonomy in the eighteenth century, burdened as it was by a great influx of new species. Linnaeus also set high informal standards in biological observation and formal standards in biological nomenclature, for which we can still be grateful today.

Between Linnaeus and the Darwinian revolution that would sweep special creation and the *scala naturae* away, there were one hundred years of 'post-Linnaean' taxonomy (although some of this occurred during his lifetime). The taxonomic flame was taken up not by Linnaeus's son or any other Swede, but by a loose affiliation of Parisians.

Nearly all immediate post-Linnaean taxonomy has a Gallic flavour, with most of the players heralding from the *Jardin des Plantes* in Paris (it was once called the *Jardin du Roi*, but this name became unfashionable after 1789). Here, in figures such as Georges-Louis Leclerc (Comte de Buffon), Michel Adanson, Antoine Laurent de Jussieu, Georges Cuvier and Jean-Baptiste

Lamarck, something approaching a natural system of classifi-cation was devised and the path for Darwin laid smooth. Two important aspects derive from these men and many others working at the time: a disdain for merely artificial systems of classification, with a sense that many characteristics should be employed in order to arrive at a natural system, and the begin-nings of evolutionary thought.

Georges-Louis Leclerc, Comte de Buffon (1707–1788)

This colourful aristocratic figure, colourfully named,* would have none of Linnaeus or any of his systems, artificial or natural. He considered the natural world to be a continuum, a view fitting in well with the *scala naturae* concept of plenitude, although his overall views on the *scala* are not entirely clear.[99] The individual organism was the most important entity to him, and it was through this notion that his very modern species concept (see p. 243) of a 'succession of individuals that can reproduce together' arose.[100] Not being particularly happy with any system of classifi-cation, he accepted only the genus as a taxon above species, and that only as a matter of convenience. He was a direct contem-porary of Linnaeus, having being born in the same year, 1707. But he outlived Linnaeus by ten years and his monumental work, *Histoire Naturelle*, was published in thirty-six volumes over the astonishingly long period of 1749 to 1788.

Leclerc's most notable achievement, beyond the grand explo-ration of the whole of the natural world that was *Histoire Naturelle*, were his thoughts on evolution. These were ill-formed and naive, being little more than the transformation of species

---

* He looks like a wedding cake in the Drouais painting of him.

into allied species due to ecological pressure, but it was a small step in the right direction. Darwin indicated just how small in the third edition of *On the Origin of Species* – 'Passing over . . . Buffon, with whose writings I am not familiar' – although he slightly improved his estimation in a later edition. The distinguished twentieth-century evolutionist Ernst Mayr was much more complimentary, describing Buffon as the 'father of evolution'.

Adanson (1727–1806) & Jussieu (1748–1836)

Buffon or no, a natural classification system was needed, and it was left largely to the Comte's fellow Frenchmen at the *Jardin* to begin to provide one. Michel Adanson and later Antoine Laurent de Jussieu took a more analytical approach to plant classification. They both looked at a large number of characteristics of each plant. De Jussieu took more notice of some than of others, a practice known as the 'subordination of characters'. For example, having two cotyledons (embryonic leaves) instead of one – i.e. the fundamental botanical distinction between dicotyledons and monocotyledons – is far more important for classification than anything associated with flowers, although generations of botanists seem to have been unaware of the fact.

As de Jussieu's predecessors had discovered, choosing the correct characters is difficult. It is only by having a comprehensive familiarity with a group of organisms that it can be done with any hope of arriving at the right classification, because the relevant character may change from one group to the next and from one taxonomic level to another. For example, Linnaeus used the character of the teeth to define the orders of mammals, but teeth were of little use in classifying the orders of reptiles or fish (although they did help in determining the *genera* of reptiles and the *species* of fish).[101] The 'natural system' for de Jussieu was nothing like what we mean by it today; for him, organisms formed a continuum within the *scala naturae*, with one species close in all things to its neighbours, so there could be no jumps, no gaps. His 'natural system' just placed together things that were similar, with the at least theoretical expectation that the gaps could one day be filled with species. Despite attempting to fit the natural world into the *scala naturae*, de Jussieu, by using many characters in his determinations – as championed by Adanson – succeeded in producing a remarkably modern classi-fication of plants, which was close to the true natural arrangement. His system for ordering plants was vastly superior to the self-consciously artificial arrangement of Linnaeus (see p. 201), and three quarters of the families he established still exist today.

Georges Cuvier (1769–1832)

Georges Cuvier took up the method of his compatriots and applied it to zoology with crucial innovation in *Le Règne Animal* of 1817–30. A friend and admirer of de Jussieu, initially he followed Linnaeus's vertebrate-dominated view of the animal kingdom,* but then started a revolution in taxonomic thought that put the vertebrates in their place and found a more natural position for everything else. The vertebrates, instead of accounting for two thirds of animals, now became just one of four sub-kingdoms. Cuvier looked at all the characteristics of animals and decided, chiefly, on the nervous system as being a class-defining character

---

* Zoology was not Linnaeus's strong suit; Ernst Mayr even suggested that his arrangement of the animals was inferior to that of Aristotle two thousand years earlier.

for his sub-kingdoms: Vertebrata (mammals, reptiles, fish), with a spinal nerve cord connected to a brain, and an internal skeleton; Articulata (insects and segmented worms), with two ventral (the opposite location to that of the vertebrates) nerve cords organised into ganglia, external skeleton and small brain; Mollusca (snails, squid), with no nerve chord but possessing distributed ganglia, nerves and a distinct brain; Radiata (sea urchins), lacking a nervous system (as far as Cuvier could see) and having radial symmetry.

This arrangement has stood the test of time remarkably well; it was certainly an enormous improvement on what went before. Ordering the animals in such a way had a profound effect on the philosophy of classification. With animals in such radically different groupings, the many intermediate species demanded by the doctrine of plenitude in the *scala naturae* could not exist. There was no point looking for an animal that was halfway between any of his four sub-kingdoms.

Cuvier's system was well received, and the blind alley of the *scala naturae* finally eradicated from biological classification. His influence on classification stretched further; his development and use of comparative anatomy and the functionality of parts of the bodies of creatures led him to the principle of the correlation of parts: 'Today comparative anatomy has reached such a point of perfection that, after inspecting a single bone, one can often determine the class, and sometimes even the genus of the animal to which it belonged,' he wrote.[102]

This has been of fundamental importance to palaeontology, where sometimes a single bone is all there is. Although an essential boon to those studying organisms of the past, in overenthusiastic hands it has occasionally resulted in some of the most extraordinary, and often amusing, mistakes in biology (see p. 47). Cuvier himself was a great fossil hunter (not immune from making odd inferences from what he found) and wrote on the subject. With this interest in what he knew to be extinct life forms, it is all the more strange that he resolutely refused to believe in evolution of any sort. His anatomical studies led him to the conclusion that there was not much you could change about an organism before it became unviable – an understandable view, even if wrong, that has been resurrected in molecular form by the modern 'intelligent design' movement.[103] Cuvier believed in 'catastrophism', the obliteration of species by large, destructive events. This theory has its own history; fashionable until it yielded to the opposite hypothesis of 'uniformitarianism' – the idea that the earth is moulded by slow, steady change – then resurgent in the latter half of the twentieth century, when mass extinctions due to errant asteroids, volcanism and similar annoyances were recognised. Of course both views are correct because the Earth is old enough for both to occur.

Jean-Baptiste Lamarck (1744–1829)

In his dismissal of evolution, Cuvier found himself in opposition to another great thinker of the time, Jean-Baptiste Lamarck. The reputation of Lamarck has suffered greatly from being placed in opposition to that of the triumphant Darwin. His theory of evolution by acquired characteristics (for example, the long neck of the giraffe would have been acquired through generation after generation of stretching to reach higher branches) was plain wrong* and a little absurd, but it was still a mechanistic biological

---

* Well, probably not entirely. Articles and papers frequently appear suggesting that sometimes it can happen.

theory to explain adaptation and diversity. Writing a little before Cuvier, Lamarck was a thoroughgoing advocate of the *scala naturae*, producing one of the most uncompromising plant 'taxonomies' honouring that paradigm ever published. It was, as one would expect, just a list, with mushrooms at one end and a complex flowering plant at the other, with little acceptance of taxa beyond species at all.

Despite his failed evolutionary theory and backward-looking taxonomy, he redeemed himself in several ways: he took Linnaeus's Insecta (insects, spiders and crustaceans) and Vermes (the waste-basket taxon, which included molluscs, worms, coral, some seaweeds and everything else Linnaeus couldn't find a class for) and placed its occupants into seven 'classes', which are still recognisable today. He also blessed us with the word 'invertebrate', for which we should be thankful. A further invention was that of the dichotomous key, the saviour of many a confused naturalist unable to identify what he has before him. The dichotomous key is a series of questions that have two possible answers, which lead (all being well) to the correct name of the species. To be frank, these can be frustrating to use (and, from personal experience, one hundred times more so to write) but they often work where all hope has been lost at trying to identify what you have found.

Finally, although Lamarck's evolutionary theory was the wrong one, it did prepare the way for Darwin and make Darwin's theory more immediately acceptable than it would otherwise have been. Lamarck and Cuvier make an odd pair: the former an evolutionist who accepted no hierarchy, and the latter having made a structured, natural hierarchy unavoidable while refusing to accept evolution. Between them they had something close to

the answer, but it was in England, not France, that the puzzle was
finally solved.

Charles Darwin (1809–1882)

> *Naturalists, as we have seen, try to arrange the*
> *species, genera, and families in each class, on what is*
> *called the Natural System. But what is meant by this*
> *system? . . . Expressions such as that famous one by*
> *Linnæus . . . that the characters do not make the genus,*
> *but that the genus gives the characters, seem to imply*
> *that some deeper bond is included in our classifications*
> *than mere resemblance. I believe that this is the case,*
> *and that community of descent – the one known cause*
> *of close similarity in organic beings – is the bond, which*
> *though observed by various degrees of modification,*
> *is partially revealed to us by our classifications.*

In *On the Origin of Species by Means of Natural Selection* in 1859,
Darwin supplied, finally and irrefutably, the answer to the question
that naturalists had fretted over, merely suspected or just plain
ignored. Living organisms could be placed in groups within groups
because of common descent. The *scala naturae* and Aristotelian
logic that had together or separately defeated attempts at a natural
classification fell to Darwin's evolutionary sword.

It is for his exposition of a natural means by which species
occur – their mechanistic evolution by natural selection acting on
variation – that Darwin is remembered, but his revolutionary
effect on taxonomy is often neglected. As Ernst Mayr wrote in
1982, echoing Alfred North Whitehead's quip about Plato and

Western philosophy: 'Theoretical discussions of evolutionary classification in the ensuing century have consisted in little more than footnotes to Darwin'.[104]

Darwin showed that perceived relationships between organisms were often real because they were based on real genealogical relationships. However, he also destroyed the notion that such overt similarity necessarily indicated relatedness. Often this is not a problem – members of the carrot family, for example, are always identifiable as such. But, by contrast, it is by no means obvious that the possession of a stem and gills in a fungal species (such as the field mushroom) fails to tell us to which of six orders it belongs (see p. 249). Modern molecular techniques have been employed to tease out these otherwise inscrutable relationships, resulting in frequent and often major reassessments of our understanding of the tree of life. The heart of the matter is that the old idea of 'type' (in its philosophical sense) was seen for what it was: an illusion. Species were transient entities, changing their form over the eons to become 'new' species, sometimes dividing into several. Typological thinking deals with permanence, and this is something that species lack. The idea on which classifications were based, similarity of form, had to be rethought to accommodate the new paradigm, which indicated that similarity of form cannot be consistently relied on in a classification.

One would have thought that having found the Holy Grail – the knowledge that relationships between organisms have a real existence – taxonomy would have enjoyed a new dawn. But the sudden burst of enlightened activity that might have been expected did not occur. Perhaps, like a solved crossword puzzle, taxonomy was of little more interest to naturalists and they

moved on to other, more challenging, or at least more interesting, endeavours. It is one thing to catalogue a library, quite
another to read the books therein. During the following decades
those who could now certainly be called scientists began to
read. The generalists of the past gave way to the specialists of
zoology, botany, mycology, bacteriology and so on. These men
and women were less interested in classifying organisms than in
simply finding out how they worked and were aided in their
studies by such things as advances in microscopy and the development of organic chemistry. Almost concurrent with the
publication of *Origin*, for example, the cell became recognised
as the basic component of organisms. But there is more to it
than that. The classifications that had emerged by the time
Darwin's ideas exploded into biology were sufficiently 'natural'
to fit in fairly accurately with the 'real' natural classification
demanded by Darwin.

I have a friend whose party trick (a long, boring one, but
impressive nevertheless) is to complete a jigsaw puzzle with all the
pieces kept face down – he fits them together in the only way they
can possibly go, and when it is finished he turns it over to reveal the
picture. You do not need to see the overall plan to complete the
picture in a jigsaw puzzle: you just have to make everything fit,
then turn it over. Darwin took the upside-down puzzle that was
taxonomy and turned it over to reveal common descent.

The long-established Linnaean hierarchical system of
kingdom, class, order etc. continued and was sometimes expanded
to fit an ever-growing number of species and groupings. New
genera were created and old ones reassessed. But there was always
something of the amateur naturalist about all this. It is generally
considered that this is where taxonomy got its bad name as poor

science – less a science, in fact, than a country craft. Where was the rigour, where were the numbers? Classification continued to be what it always had been – a matter of opinion.

Taxonomy continued nonetheless, albeit without the glamour of the eighteenth century. Common descent was there as a guide, but no more than that. Sometimes it seemed as though Darwin's taxonomic revolution had never occurred.

Phylloscopus trochiloides

CHAPTER IX

WHAT *is a* SPECIES?

IN THIS CHAPTER AND THE NEXT, I LEAVE BEHIND THE COMFORTABLE world of frock-coated naturalists, collecting baskets and quill pens and enter the modern domain of computers, DNA analysis and what I can only describe as an entire horror-fiction library of conceptual nightmares. The latter include such ideas as 'neighbour joining', 'parsimony analysis', the Genealogical Concordance Species Concept and phylogenetic systematics. This is ghastly stuff and not for the faint-hearted, but it is all fundamental to the task of classifying and naming the natural world. It is helpful at least to understand the problems that such concepts have been devised to solve, if not how they actually achieve this. I have tried to put it all as straightforwardly as possible. So, buckle down, pour a drink and, if you wish, read on.

About twenty years ago, I found a patch of exceptionally stunning-looking mushrooms. They had a bright pink stem and pale lilac gills but the cap was a brilliant yellow. I knew – because of the pink stem, the texture, a chemical test on the spores and several other factors – that it was *Mycena pura*, but *M. pura* is noted for being decidedly pink all over; certainly not the egg-yolk yellow possessed by my specimens.

I sent a couple to a respected mycologist friend, the late and much lamented John Keylock. He told me that it could be nothing other than *Mycena pura*, but if I was convinced it was a new species then I should jolly well publish. Unknown to me at the time, in 1876 a French mycologist had had the same experience and did publish, not as a new species or even a subspecies, but as a variety: *Mycena pura* var. *lutea*. In 1883, a second French mycologist decided that it was a good species and it became *Mycena lutea*, followed in 1938 by a third French mycologist, who demoted it to form, making it *Mycena pura* f. *lutea*. However, modern mycologists are very suspicious of varieties and forms, and now it is back to plain old *Mycena pura* again.

For myself, if I found it again on a foray I would say it was *Mycena pura* f. *lutea*, if for no other reason than to acknowledge that it is so very different from all the other *M. pura* we found that day. There is a separate species, *Mycena rosea*, which is just a little more pink than *M. pura*, although in the field I cannot tell it consistently from *M. pura* and I know few who can. *M. pura* was variously named by people who spent their lives studying fungi, so how could they come to such different conclusions about what constitutes separate species?

The greenish warbler, *Phylloscopus trochiloides*, inhabits the forests of central Asia circling the Tibetan Plateau.[105] The birds breed with their neighbours, producing viable and fertile offspring. However, in the far north of their range, adjacent populations refuse to have anything to do with each other. Nowhere around the Tibetan Plateau can anyone say that the species has changed from one to another, the populations grading almost imperceptibly, varying slightly in plumage and song. If this is one species, how can there be two populations that do not breed? This is an example of a rare biological phenomenon called a 'ring species'. The founder population is believed to have originated in the south and spread northwards on both sides of the Tibetan Plateau, eventually meeting in the north. The two branches evolved during their northward spread but in different ways, notably and crucially in their mating calls. Is *Phylloscopus trochiloides* one species or two? If two, where does it change from one to the other?

The North American red wolf, *Canis rufus*, which is being reintroduced to areas of the U.S. where it had become extinct, is morphologically halfway between the coyote, *C. latrans*, and the grey wolf, *C. lupus*. However, some taxonomists consider it to be a subspecies of the grey wolf and call it *C. lupus rufus*. A paper in 2011[106] revealed that in fact the red wolf is eighty per cent coyote and only twenty per cent grey wolf. So is it a distinct species, a subspecies of the grey wolf, a subspecies of the coyote or just a hybrid with no true Latin name of its own?*

The above examples scratch the surface of a longstanding and largely intractable biological conundrum known as the 'species problem'. In essence: how do you define a species? (Some would ask an even more fundamental question: do species actually exist?) The concept of what does or does not constitute a species is of enormous importance to the practice of nomenclature – how can you give a species a name if you don't know what you mean by 'species'?

---

* More fundamentally, the red wolf provides an example of 'reticulating phylogeny', when species split into two species (coyotes and red wolves have a common ancestor) then merge again to produce a third organism, which may, or may not, be a new species. The branches on the tree of life can grow back together again.

Most people would be surprised to hear that such a problem exists: everyone knows what an elder tree, a magpie, a lion or a dandelion looks like, so how can there be any debate about whether or not they are species?

Linnaeus and his contemporaries had their own species concept, although they did not recognise it as such. It was born of the notion that species are the result of special creation and that there is a 'perfect' form of each species, an idea that goes straight back to Plato. This is now called the Typological Species Concept (TSC). It defines a species as a group of individuals with certain characters in common; they are all of a 'type'.* This venerable concept, couched though it can be in the terms of an ancient philosophy, is the species concept held by most people, including taxonomists. A glance through any flora or guide book assures us that, in general, morphology determines species: if organisms look more or less the same, they constitute a species.

However, two different species can look almost identical, and organisms from the same species can look remarkably dissimilar. Just how dissimilar can be seen in the large tubeworm *Ridgeia piscesae*, which is found off the coast of the Pacific North West. It grows in two ecologically determined forms so completely unlike one another (one grey/green and smooth, the other white and woolly with bright red fronds at the end) that it was long thought to be two species. It was only after molecular analysis that the creatures were found to be a single species. Such stages or states in the life cycle of an organism are called morphs, and they sit unhappily with traditional type-based thinking.

---

* 'Type' in this context is not to be confused with the 'type' specimens (see p. 115) that are kept when a species is named. These link a specimen to a name, not to a species. Such are the fine distinctions of taxonomy.

Similarly, in mycology, there is the problem of species that exist, potentially, in two states: the sexual phase and the asexual phase. These can appear quite different, the former having a definite structure, the latter often looking more like an amorphous mould. Some reproduce only asexually, so the sexual stage is never seen in nature. For a very long time the connection between the two states was not understood, and mycologists gave them separate species names. But molecular techniques have enabled the two states to be matched, and two names (and two type specimens) for one species is not what anyone wants. At the Melbourne Congress of the ICN in 2011 it was decreed that only one name should be allowed.* Although long usage protects some names, priority usually decides which name is retained.

In addition to one species looking very much like two, there is the contrary problem of two species looking like one. So called 'cryptic' species abound in all the kingdoms. In the larger fungi, where there is often little to go on, species can be so similar that few can tell them apart. Sometimes they cannot be told apart for very good reason – they are the same species. I spent years puzzling over the difference between two large mushrooms – *Leccinum quercinum* and *L. aurantiacum* – only for the wretched things to be synonymised.

The pre-Darwinian notion of species being fixed entities contrasts sharply with the realisation that they are changeable, transient, evolutionary entities. Any perception of species being

---

* The 'mistake' was so well established by the time it was discovered, and the states so distinct, that it became a requirement under the botanical code to name both. When the ICN changed position on this in 2011, allowing only one name, it was to the disgust of several mycologists, who very nearly took to the streets.

discrete is an artefact of our own short lives. A lion may be a lion today, but it is just in the latest stage of an evolutionary process that began billions of years ago. How far back could we go before we would stop calling it a lion? How much would it have to change for us to consider calling it something else? New species, when they first split from the family tree, would appear as subspecies. When would we call them species?

Darwin understood how his theory would damage the long-held conception of what constituted a species:

> *In short we shall have to treat species in the same manner*
> *as those naturalists treat genera, who admit that genera*
> *are merely artificial combinations made for convenience.*
> *This may not be a cheering prospect, but we shall at*
> *least be freed from the vain search for the undiscovered*
> *and undiscoverable essence of the term species.*

At best, all we can say is that a group of organisms constitutes a particular species now (or, in the case of an extinct species, then).

Endless books and papers have been written about the species problem, and a variety of solutions offered. Many authors have strayed from biology into philosophy and even psychology to tackle the issue – not very helpfully, in my opinion. But perhaps I am being harsh. After all, the naming of things is a human activity, not a biological one. The human need to group and classify does not necessarily find a comfortable place in the untidy world of biology, but this is our problem, not biology's.

At the last count, there were twenty-six species concepts on offer, twenty-five more than we would like. Some are specialised concepts for the special situations encountered in certain groups, and some are general. A popular, if ultimately doomed, attempt at a general solution is the Biological Species Concept (BSC) promoted by the evolutionary biologist Ernst Mayr. It is not perfect, but at least it is comprehensible – more than can be said for some of the others.*

Ernst Mayr's definition was not exactly new, he just put it better than anyone else, and certainly more often. His idea was that a species consists of a reproductively isolated group of individuals that are able to breed together to form offspring, which can, in turn, breed successfully. This seems perfectly sensible and few would disagree with it, but biology is not sensible. The first problem with Mayr's concept is that, for any given species, it is all but impossible to discover if it fits the definition, making it useless for determining whether or not it counts as a species or not. Secondly, reproductive isolation is a hard trick to pull off and many organisms form hybrids. Around ten per cent of animal species are able to produce hybrids,[107] and many of the offspring are viable and fertile. Wolves and coyotes, as discussed, are known to have hybridised in the past, and it is with dismay that conservationists have observed them hybridising today. Plants are even less fussy. Plant hybridisation has caused many botanists to give up on the idea of BSC altogether, while entomologists and mycologists never accepted it.

The third problem is that because the BSC is based on reproductively isolated groups, it is confounded completely by the large

---

* The Least Inclusive Taxonomic Unit Species Concept, for example, is 'a taxonomic group that is diagnosable in terms of its autapomorphies, but has no fixed rank or binomial'. The Genealogical Concordance Species Concept tells us that 'Population subdivisions concordantly identified by multiple independent genetic traits constitute the population units worthy of recognition as phylogenetic taxa.' No doubt you get the idea.

number of organisms that reproduce without sex. Years ago my daughter brought home a couple of stick insects, *Carausius morosus*, in a party bag (together with the cake and chocolates), and we kept them for several years. They bred prolifically (we had the damn things crawling all over the house at one point) but were all females. This odd behaviour is not confined to stick insects; many reptiles, too, are parthenogenic (a means of reproducing asexually, from the Greek *parthenos* for 'virgin' and *genesis* for 'birth'). Such organisms do not form reproductively isolated groups. They are reproductively isolated individuals, or individual lineages.

You will be delighted to hear that I will not be describing all of the remaining twenty-five species concepts. However, three deserve a mention. Willi Hennig, of whom more shortly, devised what is called the Hennigian Species Concept. He was a purist, a character trait that can produce theories that give unrelentingly odd results of unimpeachable consistency.

Hennig was very, very fond of trees. Not the woody kind, but phylogenetic trees – which branch when one species splits into two. The point at which this split occurs (the node, if you like) he termed a 'speciation event', and at this point, he considered, two new species are created. At least one (and arguably both) of these new species is virtually indistinguishable from what it was before the speciation event and may remain so. For example, if we studied the lion for the next half a million years and found that it spawned a new species, we would have to find new names both for the lion and for the (other) new species. Furthermore, evolutionary change between speciation events is ignored, so even if our lion changed dramatically over half a million years (without dividing) it would still be the same species. Hennig's idea, unsurprisingly, has suffered a great deal of criticism. But it is at least consistent, and it nicely tackles the problem of species evolving over time.

Hennig's species concept is highly theoretical and not terribly useful to the jobbing taxonomist, something it has in common with most of the other 'concepts'. This brings us, with a sigh of relief, to two allied species concepts, the Taxonomic Species Concept and the Pragmatic Species Concept. The latter, devised by Professor David Hawksworth, puts practicality above theory by defining a species as 'a population of individuals distinguished from others by discontinuities in shared inheritable characters, and to which it is useful to give a name'. Even less bound by theory is the Taxonomic Species Concept, described by John Wilkins in his excellent review of the entire subject, *Species: A History of the Idea.* He sums up this Humpty-Dumpty definition as 'Whatever a competent taxonomist chooses to call a species'.

Modern analytical techniques have made it much easier to be 'competent' today than it was in the past. I was visiting a tax-onomist friend in his lab, and he showed me the results of a partial DNA analysis of some samples of fungi. Several already named species in the same genus were analysed at the same time, and they formed distinct clusters on a graph, as would be expected for distinct species. A putative new species fell neatly into a cluster with a known species, unequivocally telling us that it was not new. I have heard this type of analysis described as 'liquidizer taxonomy', because morphology counts for nothing; the specimen is simply mashed and analysed.

Dendrelaphis caudolineatus

CHAPTER X

The TREE of LIFE

IMAGINE THAT ONE MILLION YEARS AGO THE PROGRESS OF A PLANT population was being carefully watched and recorded by a dedicated group of robots. The plant was a member of the carrot family (Apiaceae). Its population was split in two by a new lake and interbreeding between the populations on either side became impossible. One half of the original population was happy in its environment and remained more or less unchanged over the years. The other half of the population found itself under environmental stress and started to evolve rapidly. (This is a thought experiment, so I can have it evolve as fast as I like: very fast indeed.) After one million years it evolved to look exactly like the meadow buttercup (*Ranunculus acris*) with every detail the same, from its molecular level up to an uncanny ability to determine whether or not you like butter.

Here is the crucial question: is the dramatically evolved population a member of the carrot family (Apiaceae) or is it a buttercup (in the Ranunculaceae)? There can be no doubt that it is a member of the Apiacaea, because that is what it descended from, but without reference to the careful records of some very bored robots, there is no physical evidence to prove it. In short, it is family history that counts, not morphology. Of course, a carrot could never evolve into a buttercup, but similar, if less absurd, situations have occurred countless times in nature. Fungi in particular are masters at pulling off this trick.

In the days when PowerPoint presentations were new and cool, I made one for my students. One of the slides was a highly simplified hierarchical classification of the fungi. It showed the two major divisions – Myxomycota (slime moulds) and the Eumycota (everything else) – then traced the two subdivisions that produce large fruiting bodies: the Basidiomycota (most mushrooms and bracket fungi) and the Ascomycota (morels, cup fungi, truffles, yeast). Then came the various classes, orders, families and genera. I was terribly pleased with it and my students duly impressed by my obvious depth of knowledge, mastery of technical wizardry and ability to spell difficult words.

I was due to give a lecture to a natural history society a couple of years ago, and searched out my old presentation. When I looked at the slide that showed my fungal hierarchy, my heart sank. Impressive it still was, but from top to bottom it was almost completely wrong. To be honest, it was not exactly cutting-edge at the time, reflecting an understanding of fungal relationships five or ten years old, but it was not too far from what most mycologists had accepted in those days.

Perhaps the most outdated order on my slide was the Aphyllophorales. This was a 'waste-basket' taxon, where taxonomists throw stuff that doesn't seem to belong anywhere else. Basically it was a 'to do' list. The name Aphyllophorales means literally 'without leaves', although in this context it meant 'without gills', indicating that they were not mushrooms. It also included several groups of fungi that did possess gills, so it was not a happy order right from its inception (by Carleton

Rea in the 1920s). Now it has been abandoned completely and its eight component families have been spread among existing and new orders. But even orders that had been thought of as settled had become unrecognisable since my slide was created, gaining or losing component families, splitting or being dispensed with altogether.

The familiar gills that you see every time you chop up a mushroom for tea are the structures on which the spores are produced. They are the perfect device for their function, providing a large surface area for spore production and arranged vertically so that the spores can fall freely. It is no surprise then that they have evolved independently eight times in eight major orders of fungi. This is similar to the multiple evolutionary instances of the eye, which developed independently in vertebrates, insects, squid and so on, but more impressive because, superficially at least, the gills in different orders look almost identical. However, between these eight orders it is not only the gills that appear more than once.

Just because things look similar, it does not mean they are closely related. Distinct morphologies appear again and again in different orders of fungi. For example, Agaricales contains fungi with gills, fungi with tubes in place of gills, fungi that look like corals, fungi that produce their spores inside a stomach-like structure, truffle-like fungi and fungi that appear as unimpressive fuzzy bits that grow on logs. However, these forms also appear in another order, Boletales, having evolved quite independently.

Gross morphology has proved to be a poor guide to relatedness – evolution frequently produces species that look the same because they face the same problems and produce similar organs to solve them. This is called convergent evolution, and the convergent characteristics involved are known as homoplasies (loosely translated it means 'from the same mould' or 'the same shape'). By contrast, some characteristics are held in common between species because they are inherited from a shared ancestor. The legs of a dog and those of a cat, for example, are inherited from a long-lost ancestor, which possessed a similar structure. Mammalian legs evolved only once and indicate relatedness. Such characters are called homologies ('agreeing'). However, homoplasies (which do not necessarily indicate relatedness) are considerably more common than is sometimes realised. The problem for taxonomists is: how do you tell homoplasy from homology?

In the absence of an army of ancient robots to keep an eye on things, biologists must decide on their taxonomies using specimens of extant or recently extinct organisms and maybe a few fossils if they are lucky (although not that lucky: fossils are seldom much use). Organisms have characteristics (gross morphology, microstructure, chemical signatures, DNA, habitat, behaviour), and it is on these that their taxonomies must be decided. Choosing the characters that are useful in ordering a taxonomy is extremely important and the choice will vary enormously from group to group and from rank to rank.

Historically naturalists have chosen from a basket of characters on which to base their systems – habitat, size, corolla, stamens and pistils, legs, breasts, nervous system and so on. In the case of marsupials, for example, their unusual reproductive arrangements were sufficient to separate them from placental mammals. Mycologists have used microscopic characters in the classification of fungi for more than 150 years. With the

Polygonaceae (sorrel, rhubarb), botanists must look to an array of structural details to decide on a taxonomy.

One frequent assumption among early classifiers of plants and, more latterly, animals was that the more important the character was for the life of the species, the more relevant it was to a natural classification. Darwin, himself an accomplished taxonomist (famously of barnacles), knew that this was not always the case and that often physiologically insignificant characters were better indicators of relatedness, at least at certain taxonomic levels.

Of increasing importance in taxonomy now is the analysis of organisms at the molecular level, usually by inspection of part of their DNA. (I do not wish to overstress the importance of DNA analysis. To taxonomists it is just another morphological character set, and it is often beset with problems and sometimes gives results that contradict gross morphologies and even common sense.) On the whole, changes in DNA due to mutations become more frequent the further apart organisms appear on a family tree. By DNA-matching sequences from the organisms in question, counting and comparing the number of changes (it is vastly more complicated, but this will have to do!), a phylogenetic tree – one that reflects the true phylogeny, i.e. how the organisms actually are related – can be constructed. Recent advances in these techniques have made them readily available to taxonomists, and most modern taxonomic papers present supporting results from DNA analyses.

These modern methods (and indeed more careful examinations of relationships between organisms using traditional techniques) have overturned many long-held preconceptions, and this has

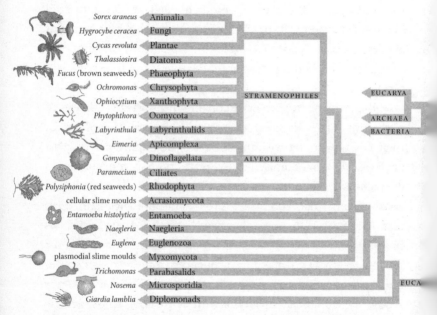

A fairly modern view of the 'tree of life', picturing sample taxa from each major group. It is humbling to note that animals, plants and fungi are consigned to a tiny corner at the top! Most of life consists of microscopic species, however many of the familiar seaweeds such as the red seaweeds and brown seaweeds are also a long, long way from the animals, plants and fungi.

had profound implications for classification and nomenclature. Existing groupings have fallen apart and new ones have been formed. Many microscopic and even some macroscopic organisms, some already known, many newly discovered, have found themselves on previously unsuspected branches of the tree of life far removed from animals, plants and fungi. The tree of life as it is seen now would be unrecognisable to Darwin; to Linnaeus it would seem absurd.

No doubt most of us can live with these considerations; some people are well aware that many (real) trees are related to smaller plants, but happy to leave fine distinctions between seaweeds or fungi to specialists. But what if we could no longer discuss reptiles unless we were happy to include all birds within that grouping, or could no longer talk consistently about fish?

Until about fifty years ago, taxonomists had been content to continue with the 'common-sense' classifications of the past. New families were discovered, unwieldy genera split and groupings moved around when relationships became better appreciated. The long-established hierarchical system of kingdom, phylum, order etc continued and was sometimes expanded to fit an ever-growing number of species and group-ings.* The evolutionary hierarchy imposed by Darwin was accepted, but determining exactly how to apply it with the limited information available at the time was difficult, and some were disinclined to take it as seriously as one might expect, content to classify according to convenience rather than principle.

* Intermediate rankings now abound, with 'superclass', 'infraclass', 'grade', 'cohort' and others filling the gaps between the familiar rankings of order, class and so on.

But during the past half century, revolutionary changes have been pronounced. For example, subsequent work on the once widely accepted classification of flowering plants laid down by Arthur Cronquist in 1981, the 'Cronquist system', has shown that three quarters of his orders contained families that should not have been there.[108] Its failing was that it took little notice of characters resulting from convergent evolution.

That such upheavals have occurred at every taxonomic rank, from kingdom* to species, means that something fundamental must be going on. While vastly improved techniques for determining species and their relationships are a major factor, the most important is that taxonomists are facing up to the implications of Darwin's theory of common descent.

The first attempt, in the early 1960s, at putting taxonomy on a more scientific footing did not, unfortunately, take common descent seriously enough, chucking out the phylogenetic baby with the traditionalist bathwater, but it did make taxonomy a little more respectable by providing it with numbers. Lots of numbers. It is called Phenetics. Linnaeus and many others thought that a natural system could only be revealed by considering as many of an organism's characters as possible. Pheneticists think so, too. The bare minimum is considered to be thirty, but a hundred or more is preferred. One writer, G. C. Steyskal,[109] believed that, with some organisms at least, no self-respecting pheneticist would try to get by with less than a thousand. Pheneticists measure, count and qualitatively assess everything.

---

* The number of accepted kingdoms is now approaching double figures and the taxonomic level of kingdom is now placed within a 'domain', of which there are now two or three.

Macro and micro morphology, chemical constituents, behaviour, habitat, DNA and more are all thrown into the mix – into a table, in fact. The more closely the characters of any two or more species (or subspecies or varieties) match, the more closely they are considered to be 'related'. It is a matter of beating a taxonomy into shape with a big stick.

Originally, as a matter of principle, all characters were considered to have equal importance. You may think that having feathers is taxonomically more important than having blue eyes, but pheneticists do not – characters were, in the terminology, 'unweighted'. The general idea was that things all work out in the end if a sufficient number of characters are considered. Unfortunately things did not, so weighted characters were allowed in some systems.

If there were complaints before that taxonomy was not 'scientific' enough, phenetics, rightly or wrongly, laid them to rest. In addition to the practice of careful observation used by taxonomists through the centuries, phenetics has a considered methodology behind it and employs mathematics. Almost invariably it requires a computer on which to run the sums, complete with a large array of analytical programs. Words and phrases such as 'phenogram', 'principal component analysis', 'cluster analysis' and 'neighbour joining' arrived on the taxonomic scene, where only the expertise of individual scientists or naturalists had appeared before.

What do you get when the computer stops humming? You get a phenogram, showing the computer's best guess at the similarities between the organisms. What you do not get, necessarily, is a phylogeny – an understanding of their true relationships. The fact that ancestral and new traits (homologies and homoplasies) are all considered equally can suggest relationships that do not truly exist. Phenetics does not produce a natural system based on common descent; it provides an artificial one based on overall resemblance. For this reason phenetics is now the poor man of taxonomy, yet it is still very useful at the level of species and below (varieties, subspecies and so on). It is only when it is applied to higher classifications that things go seriously awry.

Cladistics

The triumphant successor to phenetics is cladistics, which was invented by the German entomologist Willi Hennig. His seminal work, snappily entitled *Grundzüge einer Theorie der phylogenet-ischen Systematik*, was written in 1950, but received little attention until an English translation, *Phylogenetic Systematics*, appeared in 1966. His great contribution to taxonomy was to take phylogeny very seriously indeed. He and subsequent workers devised a philosophy, a system and a methodology for arranging the living world in a strict phylogenic hierarchy.

Few taxonomic papers that deal with more than one or two species are published nowadays without the obligatory 'cladogram' showing the relationships between the species concerned. The related 'phylogram' shows more structured relationships between larger groups. These minutely complex diagrams – sometimes you need a magnifying glass to read them – are largely impenetrable to non-specialists but of endless fasci-nation to those in the field. I have spent many a happy hour minutely scrutinising phylograms of fungal genera, whistling with pleasure or shaking my head with dismay as my preconcep-tions of relationships were either affirmed or repudiated.

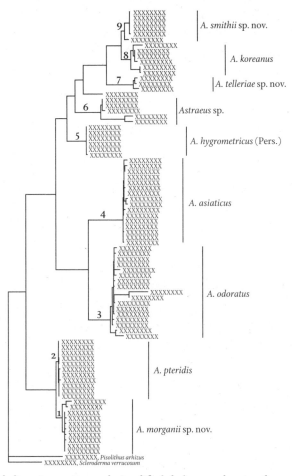

9   A. smithii sp. nov.

8   A. koreanus

7   A. telleriae sp. nov.

6   Astraeus sp.

5   A. hygrometricus (Pers.)

4   A. asiaticus

3   A. odoratus

2   A. pteridis

1   A. morganii sp. nov.

XXXXXXXX, *Pisolithus arhizus*
XXXXXXXX, *Scleroderma verrucosum*

This is, believe it or not, a much simplified cladogram showing the molecular analysis of the star-shaped fungus genus *Astraeus*. The eighty samples come from all over the world. Despite the complexity, it clearly shows how cladistic analysis has resulted in specimens falling fairly neatly into groups which can be viewed as constituting species. For clarity I have replaced the code numbers assigned to the individual specimens used in the analysis with 'xxxxxxxx'. Note the two species at the bottom of the diagram which form the outgroups. These are cousins of *Astraeus*.

A phylogram is not a classification, despite looking a bit like one, because it lacks, among other things, classes. It does, however, supply the information on which a classification can be based.

You may be wondering what cladistics does to provide a phylogeny that phenetics does not. It differentiates between ancestral characters – those that might (or might not) be present in a taxon – and those that developed later. Having four limbs (or at least indications that they once had four limbs) is not a useful characteristic with which to distinguish mammals (they all have four limbs), but the shape of the foot or arrangement of the teeth may be useful. An uncompromising attitude to common descent brings about some pretty appalling turnarounds, which also demonstrate nicely how it works. A familiar, indeed notorious, example is that of the class Aves (birds) and the orders Crocodylia (crocodiles) and Squamata (lizards and snakes). In cladistics (and in actuality), the first two are more closely related because they share derived characters – that is, characters they inherited from a common ancestor but not shared by the common ancestor of all three. Most people, upon being told that the Nile crocodile is more closely related to the robin than to the monitor lizard, would think it absurd, but it is so. These characters first appeared between the two nodes marked 'common ancestor' in the diagrams opposite.

Similarly, most of us were brought up cheerfully thinking that there was a group called 'reptiles' and a group called 'birds' – the classes Reptilia and Aves – but cladistics will have none of this. It does not allow a class (or any other taxonomic level) that excludes anything from the tree above the point at which they diverged from a common ancestor. In other words, if it requires two cuts

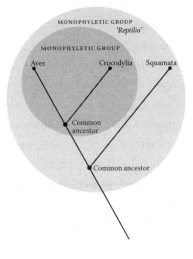

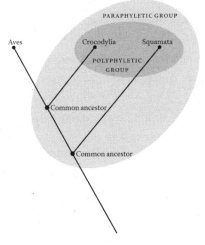

Here are two depictions of the same phylogram, indicating relationships between taxa. The top diagram shows monophyletic groups (clades); the other shows groups that are not 'natural' (i.e. not clades), but which may be considered as groups for convenience.

with pruning shears to remove a 'group', it is not a permissible group. If you want to stick with the traditional class Reptilia, then birds must be included. In the jargon, Reptilia must be a monophyletic group – descended from one common ancestor. Such a group is called a 'clade' and, as illustrated on the previous page, it is defined as everything that descended from a single node and includes the node itself. There are two clades shown in the top drawing enclosed by circles. If the birds are excluded, as in the bottom drawing, it is not a clade (monophyletic group), it is a paraphyletic group. Also not qualifying as clades are polyphyletic groups – groupings that exclude the common ancestor.

Clades can exist at all taxonomic levels, from domain to species and below, although they do not necessarily correspond to existing classes, orders, genera etc. Unfortunately the bird/reptile problem is worse than I have shown. If one considers it proper to place birds among the reptiles, it becomes difficult to argue that one should not do the same with mammals, since they arose (much earlier) from a reptilian ancestor. This conundrum is, however, trivial compared with what has happened to the fish.

Fish form no coherent taxonomic group at all. Pisces, the class beloved of Artedi, no longer exists. Like birds, fish are a paraphyletic group because of all the things we should include as fish but don't. Shrews, for example. There are nearly twenty classes of fish: ray-finned fish, armoured fish, lobe-finned fish, hagfish and so on. 'Class' is a high taxonomic rank, so these fish are only distantly related to one another. Among the classes is the Sarcopterygii (lobe-finned fish), which numbers among its subclasses the Actinistia (coelacanths), the Dipnoi (lungfish) and the Tetrapodomorpha (tetrapods). Tetrapods include frogs, salamanders, cats, dogs and, of course, us. I went to a lecture a while ago on the evolutionary origins of tetrapods, and cornered the professor afterwards. 'What does it make us?' I asked. 'We,' he responded, 'are just highly specialised fish.' This startling revelation means that we are more closely related to the coelacanth than the coelacanth is to the herring.

I will not go deeply into the methodology of cladistics, lest I drown in its dark conceptual waters, floundering among symplesiomorphies, parsimony algorithms and incongruence length difference tests. I must, however, dip in a toe. Like phenetics, it records characters (although unlike phenetics, they are a few selected characters) and enters them into a table. Ideally the data should be in the form of a 'yes' or a 'no', represented by ones and zeros (the computer's favourite numbers), although in molecular work the data come in fours, corresponding to guanine, adenine, thymine and cytosine, the coding components of DNA. Also in the table will be data from an 'outgroup' – a taxon that is closely related to the group being examined but not in that group. Usually it is a 'sister group', for example, Aves is a sister group of Crocodylia. The purpose of the outgroup is effectively to subtract many of the plesiomorphies (characters inherited from the ancestor of both the group under study and the outgroup), leaving just the new characters that have developed in the study group. This tackles the weakness in phenetics, whereby everything is taken into account. The members of the study group could be arranged in a vast number of different family trees, only one of which is the true tree. Computer programmes will generate a large number of these and examine them; the one that requires the least number of steps (parsimony analysis) is considered to be the most likely and is usually the one that is accepted and published. The principle of parsimony is central to cladistics. It is more likely that two species within a group possess a novel character if they inherited it from a common ancestor rather than both having developed it independently.

There are a few technical issues with cladistics. Convergent evolution can still be a problem, although using a large enough number of characters in the dataset can overcome this. Selecting an outgroup is usually a headache, and considered impossible in many disciplines, such as mycology. In principle almost anything will serve, but the closer it is to the ingroup, the more detailed will be the result. In studying rodent taxonomy, for example, it would be sensible to use the Lagomorpha (hares and rabbits) as the outgroup, because the ancestral traits are fairly obvious (legs, fur, vertebra etc), but ancestral traits in fungi are not so easy to determine and the resolution (accuracy) of the result may not be as good as it might be.

Perhaps the most serious criticism is that cladistics only ever deals with a branching system, even though it is well known that lineages do not just branch – sometimes the branches grow back together, in a process called reticulating evolution (see p. 239n). Hybridisation, as far as cladistics is concerned, does not exist. This is an enormous obstacle to the acceptance of cladistics in botany, where hybridisation is very common.

A major problem is that established taxonomic ranks do not fit well into the complex branching of the true phylogenies that cladistics reveals. Today one is likely to find that a grouping is a named, or even unnamed, clade with no obvious taxonomic ranking (noted as 'unranked'). Traditional rankings of family, order and so on can find themselves at inconsistent levels on the cladistic tree.

∽∞∽

I have not met many taxonomists who have taken cladistics truly to their hearts. They consider it too severe a paradigm in its pure form and choose from it the parts that suit their purpose. A letter to the journal *Taxon* in 2005 from 140 biologists from every corner of the world (the signature is two and a half times the length of the letter) pleaded that 'Paraphyletic taxa should be accepted'.[110] It also rehearsed frequent complaints about the chaos that close adherence to cladistics would wreak on taxonomy.

Despite the problems and the resistance of some scientists, cladistics pervades modern taxonomy, its methodology accepted and employed throughout the disciplines. It has taken taxonomy from being an art to what it deserves to be: a science.

The Phylocode

At around the turn of the century, a group of academics optimistically established the Phylocode, a system of rules that took an uncompromisingly cladistic view of taxonomy. Ranks would cease to exist, leaving taxonomists free to name clades as it became convenient to do so, but they would have to *be* clades. The ranks themselves would be dispensed with, although not necessarily the names of the ranks. *Sorex araneus* would likely be in the clade Sorex, for example, but it would not be in the genus *Sorex*.

Fortunately most cladists are content with the existing codes, happy to use them to name new clades when they consider it necessary. Given what taxonomists have told me, the letter to the editor of *Taxon* and a thousand other publications decrying the wholesale application of cladistics, I think it is fair to say that few proposals in scientific history have received as poor a reception as that of the Phylocode.

Hygrocybe pratensis

EPILOGUE

WRITING A HISTORY OF NOMENCLATURE AND TAXONOMY IS LIKE writing a history of the Second World War in May 1944. You can see where things are going but there is little certainty. I am, however, sure that Latin names as we have known them since the eighteenth century will continue unchanged. They are simply too useful to discard, and no sensible replacement is even imaginable, let alone practical.

If there is likely to be a problem with them, it is that many of those we know and love will have to be dispensed with, or at least used in a more restricted sense. Molecular and other techniques have begun dramatically to split familiar species and genera, while occasionally lumping others. This has been going on for years, but the new toys available to taxonomists have made wholesale reassessments of taxa all too easy, and the confusion that Linnaeus sought to avoid threatens us again. One academic I spoke to even complained that taxonomy had become something of a liability, distracting scientists from more important issues.

I particularly admire the mushrooms in the genus *Hygrocybe*, known as the waxcaps. They are stunningly beautiful, with bright colours from red to yellow to green and even pink. Molecular work has been done on these, and some species seem set to split into two or more, based mostly on the fine genetic differences that are frequently found in localised populations. Often there is little or nothing to tell them apart visually, it is just that the existing species will fall into two or more genotypic groups. The upshot of this is that if a species was named originally for a type specimen found in France, the English may find themselves with a new species and a new name for those found in England, and the Italians may be faced with a third new species and name. All three look almost exactly the same.

I do not think there is anything that can be done about this, and it is, after all, up to scientists how they arrange their classifications. But if endless splitting is to be the fate of very many species, the average naturalist or conservationist may effectively go on strike and ignore the new names.

This already happens with many cryptic species (those that are very difficult to tell apart), with the use of the term *sensu lato* (broad sense). For example, the nematode worm *Xiphinema americanum* was named, obviously, as a single species, then split into seven, then fifteen, then twenty-five species, all but one with a new name. However, it (or they) are generally spoken of as *Xiphinema americanum sensu lato*, although one of them is *Xiphinema americanum sensu stricta* (strict sense). It would be a great pity if we had to face this irritation with well-known species with which everyone is perfectly content.

For my part, I will continue to use and enjoy the names we have and those that are yet to be conceived. If a name changes, I will tolerate the annoyance as I have always done, while appreciating the new knowledge that has required the change. At least if the meadow waxcap, *Hygrocybe pratensis*, changes its name yet again, it may well revert to the one I learned forty years ago. Welcome home *Hygrophorus pratensis*.

For the ordinary naturalist taking an afternoon stroll, the world of taxonomy, the world of names, is little different from that known by Linnaeus. I am content to use the names I know and to learn those that I do not. The joy of discovering *Drosera* and its lovely Latin binomial has never left me and I find the same pleasure every time I encounter a name that is unfamiliar. To me, and to everyone whose curiosity knows no sensible bounds, names are not mere appellations, but things in themselves. Each has its own history and revealing etymology. Somewhere on a dusty shelf there will a corresponding dried or bottled specimen to which it is forever linked, and there is, or once was, an author who quite possibly loved his creation. Mythologies often speak of the power of names, of the sense of possession that comes with knowing the real name of a person or object. While the shrew may answer to a hundred names, it has only one that is true. It is *Sorex araneus*.

GLOSSARY

Artificial Classification: Classification used for convenience (alphabetical, by size, by habitat, by usage etc). cf. **Natural classification**.

Author: Person who devises or changes the name of a **species**, **genus** etc. More generally, the creator of any **nomenclatural act**.

Autonym: Name created automatically as the result of some more general **taxonomic** change. For example, a new **binomial** created when a **genus** name changes.

Binomen: Two-part name of a **species** (zoology).

Binomial: Two-part name of a **species** (botany, mycology, phycology).

Circumscription: **Taxonomic** view. For example, the view that particular **genera** belong in a **particular** family is a circumscription of that family. Views, of course, may differ.

Citation: Name of the **author** or authors of a particular **taxon**. Conventionally a citation is given after the name of a **species** or **genus**.

Clade: Any branch of the **tree of life** that can be removed by a single cut. It will and must include the common ancestor.

Cladistics: Study and creation of classifications based on true familial relationships between **taxa**.

Cladogram: Branching 'tree' diagram that displays familial relationships between **taxa** derived by **cladistic** methods.

Class: A major taxonomic rank between phylum and order.

Domain: Highest **taxonomic** level apart from 'life'. At present, there are three domains: Bacteria, Archaea, Eukaryota, the latter distinguished by having a nucleus.

Family: Major **taxonomic rank** between **order** and **genus**.

Generic name: First part of a **binomial**, denoting the **genus** to which the **species** belongs.

Genus: The major **taxonomic rank** above **species** and below **family**. Plural: **genera**.

Heterotypic synonymy: **Synonymy** resulting from more than one single name and **type** specimen being applied to a single **species**.

Holotype: Single **type** specimen denoted by the **author**.

Homonymy: When the same name is applied to two **taxa** at the same **taxonomic rank**. Most often, the same name given to two different **genera** or two different **species**.

Homotypic synonymy: **Synonymy** resulting from two names being applied to a single **type** specimen.

Hybrid: Offspring of two (or more) **species**. It may or may not be capable of reproduction.

Junior: In **homonymy** and **synonymy**, the more recent name for a **taxon**. Gives way to **senior homonym** or **synonym**.

Kingdom: Major **taxonomic rank** between **domain** and **phylum**.

Lectotype: Specimen from original material designated as a **type** by a later **author**.

Morphology: Shape, colour, gross structure, microstructure etc of an organism.

Natural classification: Classification deemed to be based on the true arrangement of **taxa**. Many have been proposed or accepted, but there can only be one: that of descent with modification in a family tree. cf. **Artificial classification**.

Neotype: **Type** specimen designated to replace a type that is lost or that never existed.

Nomenclatural act: Any published writing that affects the nomenclatural status of a scientific name for **taxa**.

Nomenclature: The naming of living organisms.

Order: Major **taxonomic rank** between **class** and **family**.

Phenetics: Arrangement and sometimes classification of **taxa** by comparing a large number of characters.

Phylogenetic systematic: See **Cladistics**.

Phylogeny: Familial or evolutionary history of **taxa**.

Phylogram: See **Cladogram**.

Phylum or Division: Major **taxonomic rank** between **kingdom** and **class**. Phylum is preferred term.

Principle of priority: Rule which states that the earliest name given is generally the correct name for a **taxon**.

Senior: In **synonymy** and **homonymy**, the earliest and generally correct name for a **taxon**. cf. **Junior**

Species: Subdivision of a **genus**. Any group of closely related organisms that are sufficiently alike to warrant a **species** name (although see Chapter IX).

Specific epithet: Second part of a **binomial**, denoting which **species** within a **genus** the whole name applies to.

Subspecies: **Taxonomic rank** immediately below **species**. Identifiable because it has two **specific epithets**.

Synonymy: When two or more names are applied to the same **taxon**.

Syntype: Any single specimen in a collection that is listed in the description where there is more than one specimen and no holotype has been designated.

Tautonym: Repetition of a word to form a binomial or trinomial, e.g. *Gorilla gorilla.* Zoology only.

Taxon: Any named or unnamed group of organisms. A **species** is a taxon consisting of individuals; a **family** is a taxon consisting of **genera.** Plural: taxa.

Taxonomic rank: Named or unnamed position in biological classification, e.g. **genus**, **class**, **phylum**.

Taxonomy: The naming and classification of the living world.

Tree of life: Branching hierarchy showing the evolutionary development of **taxa**.

Trinomen: Three-part name for a **subspecies** (zoology).

Trinomial: Three-part name for a **subspecies** (botany, mycology, phycology).

Trivial name: Alternative term for **specific epithet**. Originally used by Linnaeus to distinguish the single word from the often longer specific name.

Type: Specimen, drawing or photograph used to connect a physical entity to the name of a **species** or **genus**.

Typification: Act of designating a **type**.

NOTES TO THE TEXT

1   G. Grigson, *The Englishman's Flora* (Helicon, 1996).

2   britmycolsoc.org.uk/library/english-names/

3   A. Doyle 'Wrong names for fish seen complicating conservation', reuters.com (25 June 2008).

4   J. G. Wood (ed.), *Animal Kingdom* (1870).

5   H. H. T. Jackson, *Mammals of Wisconsin* (University of Wisconsin Press, 1961).

6   E. Figueiredo and G. F. Smith, 'What's in a name: epithets in Aloe L. (Asphodelaceae) and what to call the next new species', *Bradleya* 28 (2010).

7   D. W. Minter *et al.*, '*Zeus olympius*, gen. et sp. nov. and *Nectria ganymede* sp. nov. from Mount Olympus, Greece', *Transactions of the British Mycological Society* 88 (1987).

8   *Edinburgh New Philosophical Journal*, Volume 3 (1827).

9   J. Watson *et al.*, 'A revision of the English Wealden Flora, III: Czekanowskiales, Ginkgoales & allied Coniferales', *Bulletin of the Natural History Museum*, Geology series 57, part 1 (2001).

10 J. E. Bond, 'Phylogenetic treatment and taxonomic revision of the trapdoor spider genus *Aptostichus* Simon (Araneae, Mygalomorphae, Euctenizidae)', *ZooKeys* 252 (2012).

11 M. Delcheva *et al.*, 'On the taxonomical identity of *Abies alba*

ssp. borisii-regis (Mattf.) Koz. Et Andr. – morphometry, flavo-noids and chorology in Bulgaria', *Botanica Serbica* 34 (2) (2010).

12 T. L. Erwin, 'The Beetle Family Carabidae of Costa Rica: Twenty-nine new species of Costa Rican *Agra* Fabricius 1801 (Coleoptera: Carabidae, Lebiini, Agrina)', *Zootaxa* 119 (2002).

13 curioustaxonomy.net

14 R. Hoser, 'What's in a species name?', *Crocodilian* 2 (7) (2000).

15 *New Scientist*, 13 May 1982.

16 S. Connor, *The Book of Skin* (Reaktion, 2003).

17 T. L. Mitchell, *Three Expeditions into the Interior of Eastern Australia*, volume 2 (1838).

18 R. Dunn, *Every Living Thing: Man's Obsessive Quest to Catalog Life, from Nanobacteria to New Monkeys* (HarperCollins 2009).

19 B. V. Brown *et al.*, *Manual of Central American Diptera*, Volume 1 (National Research Council of Canada, 2009).

20 D. J. Patterson and J. Larsen, *The Biology of Free-living Hetero-trophic Flagellates* (Clarendon Press, 1991).

21 C. Linnaeus, *Linnaeus's Öland and Gotland Journey*, translated by M. Åsberg and W. Stearn (Linnean Society, 1973).

22 H. Wolrath *et al.*, 'Trimethylamine content in vaginal secretion and its relation to bacterial vaginosis', *APMIS: acta pathologica, microbiologica, et immunologica Scandinavica* 110, part 11 (2002).

23 T. Bailey, *Miraculum Naturae: Venus's Flytrap* (Trafford Publishing, 2008).

24 P. Fantz, 'Nomenclatural Note on the Genus *Clitoria* for the Flora North American Project' *Castanea: Journal of the Southern Appalachian Botanical Society* 65 (2) (2000).

25 M. L. Berenbaum, *Buzzwords: A Scientist Muses on Sex, Bugs and Rock 'n' Roll* (Joseph Henry Press, 2000).

26 W. Stearn, *Botanical Latin* (David and Charles, 1973).

27 J. Mavarez, M. Linares, 'Homoploid hybrid speciation in animals', *Molecular Ecology* 17, no. 19 (2008).

28 'Taxa that are believed to be of hybrid origin need not be designated as nothotaxa', *International Code of Botanical Nomenclature* H.3.3, note 1.

29 P. G. Parkinson, 'The International Code of Botanical Nomenclature: An Historical Review and Bibliography', *Tane* 21, 1975.

30 A. Sinha *et al.*, '*Macaca munzala*: A New Species from Western Arunachal Pradesh, Northeastern India', *International Journal of Primatology* 26, no. 4 (2005).

31 U. Eggli and B. E. Leuenberger, 'Type specimens of Cactaceae names in the Berlin Herbarium (B)', *Willdenowia* 38 (2008).

32 A. C. Pont, 'The Linnaean Species of the Families Fanniidae, Anthomyiidae and Muscidae (Insecta: Diptera)', *Biological Journal of the Linnean Society* 15, no. 2 (1981).

33 P. M. Jørgensen *et al.*, 'Linnaean Lichen names and their typification', *Botanical Journal of the Linnean Society* 115, no. 4 (1994).

34 R. Nourish and R. W. A. Oliver, 'Chemical studies on some lichens in the Linnaean Herbarium and lectotypification of *Lichen rangiferinus* L. (em. Ach.)', *Biological Journal of the Linnean Society* 6, issue 3 (1974).

35 D. L. Hawksworth, 'Terms used in bionomenclature: the naming of organisms (and plant communities)', bionomenclature-glossary.gbif.org (2010).

36 As described by the American entomologist Neal Evenhuis.

37 A. Ovchinnikova *et al.*, 'Taxonomy of cultivated potatoes (Solanum section Petota: Solanaceae)', *Botanical Journal of the Linnean Society* 165, issue 2 (2011).

38 S. Redhead, *et al.*, '*Coprinus* Persoon and the disposition of *Coprinus* species *sensu lato*', *Taxon* 50 (2001).

39   B. Maslin, 'Proposed Name Changes in Acacia', *Australian Plants Online* (2003).

40   G. Moore, 'The handling of the proposal to conserve the name Acacia at the 17th International Botanical Congress – an attempt at minority rule', *Bothalia* 37, 1 (2007).

41   L. Warren, *Constantine Samuel Rafinesque, A Voice in the American Wilderness* (University of Kentucky Press, 2004).

42   Ainsworth Transaction BMS.

43   G. Singh, *Plant Systematics: An Integrated Approach* (Science Pub Inc, 2004).

44   ibid.

45   solgenomics.net/organism/1/view/

46   R. Thompson, *The Assyrian Herbal* (Luzac & Co, 1924).

47   W. Sneader, *Drug Discovery: a History* (Wiley, 2005).

48   A. Luch, *Molecular, Clinical and Environmental Toxicology* (Springer, 2010).

49   Encyclopaedia Britannica.

50   M. Castlemen, *The Healing Herbs* (Bantam, 1995).

51   E. Kremers, *Kremer's and Urdang's History of Pharmacy* (Lippincott, 1976).

52   J. F. Nunn, *Ancient Egyptian Medicine* (British Museum Press, 1996).

53   Aristotle, the works of Aristotle translated into English, Vol. IV *Historia Animalium*, translated by D'A. W. Thompson (Clarendon Press, 1910).

54   A. T. Johnson and H. A. Smith, *Plant Names Simplified* (Old Pond, 2008). thepoisongarden.co.uk/atoz/aristolochia_clematitis.htm

55   A. Pavord, *The Naming of Names* (Bloomsbury, 2005).

56   W. Stearn, *Botanical Latin* (David and Charles, 1973).

57 E. L. Greene, *Landmarks of Botanical History*, edited by F. N. Egerton (Stanford University Press, 1983).

58 ibid.

59 C. Anthon, *A Classical Dictionary* (Harper & Bros, 1848).

60 Pavord, *Naming of Names*.

61 *Pliny, the Elder, Natural History*, introduction by A. T. Grafton, translated by H. Rackham (Folio Society, 2012).

62 C. Nicolet, *Space, Geography, and Politics in the Early Roman Empire* (University of Michigan Press, 1990).

63 R. Southey, *The Doctor*, edited by J. W. Warter (Longman 1853).

64 H. J. Rose, T. Wright, *New General Biographical Dictionary*, Volume 5 (Nabu Press, 2011).

65 R. Rubenstein *Aristotle's Children: How Christians, Muslims, and Jews Rediscovered Ancient Wisdom and Illuminated the Dark Ages* (Harcourt Brace International, 2003).

66 Pavord, *Naming of Names*.

67 Greene, *Landmarks*.

68 ibid.

69 Pavord, *Naming of Names*.

70 C. Knight, *The English Encyclopaedia*, 3rd Division, Biography (Bradbury, Evans & Co, 1867).

71 B. W. Ogilvie, *The Science of Describing: Natural History in Renaissance Europe* (University of Chicago Press, 2006).

72 E. Mayr, *The Growth of Biological Thought* (Belknap Press, 1982).

73 ibid.

74 ibid.

75 S. A. Baldwin, *John Ray, Essex Naturalist* (Baldwin's Books, 1986).

76 C. E. Raven, *John Ray, Naturalist, His Life and Works* (Cambridge University Press, 1942).

77 J. Ray, *Historia Plantarum* (Londini, 1686–1704).

78 Mayr, *Biological Thought*.

79 H. Hallam, *Introduction to the Literature of Europe in the 15th, 16th and 17th Century* (John Murray, 1837).

80 Linné Online, Linnaeus.uu.se/online

81 G. Cuvier, *Cuvier's Animal Kingdom* (Orr & Co, 1849).

82 W. Blunt, *The Compleat Naturalist: a Life of Linnaeus* (Collins, 1971).

83 *The Linnaean Correspondence*, linnaeus.c18.net/Doc/lbio.php

84 V. H. Heywood, *Biology and Chemistry of the Umbelliferae* (Botanical Journal of the Linnean Society. Supplements / Linnean Society of London) (Academic Press Inc, 1971).

85 B. Gardiner, 'From the Archives', *The Linnean* 18 (2002).

86 C. Linnaeus, *A Tour in Lapland*, translated by J. E. Smith (British Library, 2011).

87 C. Linnaeus, *Musa Cliffortiana – Clifford's Banana Plant*, translated by S. Freer, *Regnum Vegetabile* 148 (Koeltz Scientific Books, 2007).

88 Linnaeus, *Musa Cliffortiana*.

89 E. E. Cheesman, 'Classification of the Bananas. II. The genus L. Musa.', *Kew Bulletin* 2(2) (1947).

90 C. Linnaeus, *Philosophia Botanica*, translated by S. Freer (Oxford University Press, 2003).

91 Linné Online, linnaeus.uu.se/online/animal/2_1.html

92 Linnaeus, *Philosophia Botanica*.

93 ibid.

94 ibid.

95 J. M. Diamond, *The Third Chimpanzee* (HarperCollins, 1992).

96 T. Frängsmyr (ed.), *Linnaeus: the Man and His Work* (California University Press, 1983).

97 ibid.

98   M. A. Woodley, 'Is *Homo sapiens* polytypic? Human taxonomic diversity and its implications' *Medical Hypotheses* (2009).

99   Mayr, *Biological Thought*.

100 ibid.

101 ibid.

102 M. J. S. Rudwick, *Georges Cuvier, Fossil Bones and Geological Catastrophes* (University of Chicago Press, 1997).

103 M. J. Behe, *Darwin's Black Box* (Simon & Schuster, 2006).

104 Mayr, *Biological Thought*.

105 D. E. Irwin, 'Song variation in an avian ring species', *Evolution* 54, issue 3 (2000).

106 B. M. vonHoldt *et al.*, 'A genome-wide perspective on the evolutionary history of enigmatic wolf-like canids', *Genome Research* 21 (2011).

107 K. Schwenk *et al.*, *Hybridization in Animals: Extent, Processes and Evolutionary Impact* (Royal Society Publishing, 2008).

108 Angiosperm phylogeny website, mobot.org/mobot/research/apweb/

109 T. Stuessy, *Plant Taxonomy* (Columbia University Press, 2009).

110 I. Nordal and B. Stedje, 'Paraphyletic taxa should be accepted', *Taxon* 54, 1 (2005).

BIBLIOGRAPHY

Aristotle, the works of Aristotle translated into English, Vol. IV, *Historia Animalium*, translated by D'A. W. Thompson (Clarendon Press, 1910).

Bailey, T., *Miraculum Naturae: Venus's Flytrap* (Trafford Publishing, 2008).

Bernhardt, P., *Gods and Goddesses in the Garden* (Rutgers University Press, 2008).

Blamey, M. and R. Fitter, *The Wild Flowers of Britain and Northern Europe* (Collins, 1979).

Blunt, W., *The Compleat Naturalist: a Life of Linnaeus* (Collins, 1971).

Brown, R. W., *Composition of Scientific Words* (Smithsonian Books, 2000).

Gledhill, D., *The Names of Plants* (Cambridge University Press, 1985).

Greene, E. L., *Landmarks of Botanical History*, edited by F. N. Egerton (Stanford University Press, 1983).

Grigson, G., *The Englishman's Flora* (Helicon, 1996).

International Commission on Zoological Nomenclature, International Code of Zoological Nomenclature (ICZN, 1999).

Jeffrey, C., *An Introduction to Plant Taxonomy* (Cambridge University Press, 1982).

Linnaeus, C., *A Tour in Lapland*, translated by J. E. Smith (British Library, 2011).

—*Linnaeus's Öland and Gotland Journey*, translated by M. Åsberg and W. Stearn (Linnean Society, 1973).

—*Philosophia Botanica*, translated by S. Freer (Oxford University Press, 2003).

Mayr, E., *The Growth of Biological Thought* (Belknap Press, 1982).

McNeill, J., *International Code of Nomenclature for Algae, Fungi and Plants* (Koeltz Scientific Books, 2012).

Ogilvie, B. W., *The Science of Describing: Natural History in Renaissance Europe* (University of Chicago Press, 2006).

Pavord, A., *The Naming of Names* (Bloomsbury, 2005).

Pliny, the Elder, Natural History, introduction by A. T. Grafton, translated by H. Rackham (Folio Society, 2012).

Raven, C. E., *John Ray, Naturalist, His Life and Works* (Cambridge University Press, 1942).

Savory, T. H., *Naming the Living World, an Introduction to the Principles of Biological Nomenclature* (English Universities Press, 1962).

Stace, C., *New Flora of the British Isles* (Cambridge University Press, 2010).

Stearn, W., *Botanical Latin* (David and Charles, 1973).

—*Stearn's Dictionary of Plant Names for Gardeners* (Cassell, 1992).

Theophrastus, De Causis Plantarum, edited and translated by B. Einarson and G. Link (Loeb Classical Library 1989–90).

Theophrastus and F. Wimmer, *Historia Plantarum* (Kessinger Publishing, 2010).

Tudge, C., *The Variety of Life: A Survey and a Celebration of All the Creatures That Have Ever Lived* (Oxford University Press, 2000).

Turland, N., *The Code Decoded: a User's Guide to the International Code of Nomenclature for Algae, Fungi and Plants* (Koeltz Scientific Books, 2013).

White, T. H., *The Book of Beasts, Being a Translation from a Latin Bestiary of the Twelfth Century* (Cape, 1954).

Wilkins, J. S., *Species: A History of the Idea* (University of California Press, 2009).

Winston, Judith E., *Describing Species* (Columbia University Press, 1999).

The authors and publishers acknowledge the following permissions to reprint copyright material:

The quotations on pages 181–2 are taken from *The Compleat Naturalist* by Wilfrid Blunt, published by Frances Lincoln Ltd., 2002, and are reproduced by permission of Frances Lincoln Ltd. and the Curtis Brown Group Ltd. on behalf of the Estate of Wilfrid Blunt. Copyright © the Beneficiaries of the Literary Estate of Wilfrid Blunt 1971

The quotations on pages 206 and 209 are taken from *Philosophica Botanica* by Linnaeus, translated by Freer (2003), and are reproduced by permission of Oxford University Press.

The quotations on pages 191 and 193 are taken from *Musa Cliffortiana – Clifford's Banana Plant* by Linnaeus, translated by Freer (2007). Reproduced by permission of *Regnum Vegetabile* and Frederica Freer.

The diagram on page 252 is based on a diagram on pages 140–1 of *The Variety of Life* by Colin Tudge (2006) and the data in a diagram in *Science* magazine 290, 972 (2000). Reproduced by permission of Oxford University Press and *Science*.

ACKNOWLEDGEMENTS

It is impossible to write any book of this type without help, and I have been most fortunate in finding a large number of erudite, charming and, just as important, willing, specialists to provide it. The assistance they have given me has been unstinting and I am extremely grateful to them all. Foremost among my many advisors on matters technical is David Hawksworth. He has been generous in his suggestions and agreeably ruthless in saving me the embarrassment of the many howlers that appeared in early drafts. I also wish to thank Ole Seberg for guiding me through the opaque waters that are cladistics. Greg Kenicer, Juliet Brodie, Richard Fortey, Stewart McPherson, Tim Bailey, Graham Elmes, Phil Sterling, Bryan Edwards, Jenny Clack, Craig Rudman, Gavin Prideaux, Mike Richardson, Max Coleman, Mike Gardner, Roy Watling, John McNeill, Neville Kilkenny, Alan Hills and the staff of the Linnean Library all provided me with advice and stories. They are, in a way, the heroes of this book, for these are the mycologists, zoologists, botanists, phycologists and academics who use and often devise Latin names.

Thank you to Polly Winsor and Staffan Mueller-Wille for explaining the vagaries of Linnaean philosophical thought to me.

Many other people, notably Pip Taylor, have offered stories and advice and I thank them all.

I am grateful to Eugenio Donadoni who played the part of my tutor of old, Dr Parker, in correcting my terrible Latin. If there is still something wrong with it, blame me, not Eugenio. I am a poor pupil.

The Bloomsbury team have tolerated me with undeserved patience and kindness, and shown absolute professionalism in their determination to make this book as good as possible. Thank you to Richard Atkinson for having faith, Natalie Bellos for guiding the whole project and reading endless drafts (well, they turned out to be drafts), Alison Glossop for her detailed attention to the proofs and Rachael Oakden who re-ordered my jumbled text with great sensitivity, corrected mistakes that still keep me awake at nights and allowed so many of my jokes. I hope Rachael has recovered from the stories contained in the section on rude Latin names. Thank you to Steve Cox for his careful work on the pages and in particular for a bright idea which had passed me by but now finds itself in the book. I am delighted with the lovely, intricate drawings that grace the beginning of each chapter. These are the work of the highly talented Lizzie Harper. Thank you Lizzie.

I would like to thank Rob Love, Antony Topping and Hugh Fearnley-Whittingstall for giving me the chance to do what I now do. Gratitude is also due to my agent, Gordon Wise of Curtis Brown, for his invaluable advice and tireless encouragement.

INDEX

monocotyledons and dicoty-
  ledons, 176
Monson, Lady Anne, 36
Monsonia, 36
Moraea, Sarah Elizabeth, 196–7,
  220
Morchella esculenta (common
  morel), 79, 156
Mormoops megalophylla
  (leaf-chinned bat), 41
morphs, 240
mosses, naming of, 131
Motacilla cinerea patriciae
  (wagtail), 35
Mount Olympus, 26
Mozartella (wasp), 37
mullet, price of, 163n
Musa, Antonius, 194
Musa (bananas), 194–5
Musca (flies), 118
Mussaenda epiphytica, 45
Mutinus caninus (dog stinkhorn),
  60
Mycena
  M. *haematopus*, xii
  M. *pura*, 237–8
  M. *rosea*, 238
Myosotis (forget-me-nots), 162
Myrmica gracillima, 35–6
Myrrhis odorata (sweet cicely), 162
mythological names, 24–30
Myxomycota (slime moulds), 248

Nabokov, Vladimir, 31
Naja naja (Indian cobra), 138
Narcissus, 30, 159
Natrix natrix (grass snake), 138
Nature (journal), 50
Nectria cinnabarina (coral-spot
  fungus), 26

Neomys fodiens (water shrew),
  20
Nerocila, 69
Nessiteras rhombopteryx (Loch
  Ness monster), 50
Ng, Peter, 72
Nicomachus, 146
Nineveh, 143
Niptus, 50
nomenclatural acts, 97–8, 220
nomina conservanda, 73, 134–6
Norasaphus monroeae, 38
Nothobaccharis, 89
Nothopanax, 89
Notiocryptorrhynchus, 75
nounal names, 22–4
Nyctereutes procyonoides
  (nocturnal raccoon dog), 84

Oceanodroma, 28
Oedipina complex, 30
Oedipus rex, 30
Oenanthe (water-dropworts), 129,
  159
  O. *crocata* (hemlock water-
    dropwort), 142
Oenanthe (wheatears), 129
  O. *oenanthe*, 11, 138
Ofcookogona (millipedes), 77
Oides (beetles), 84
Olea
  O. *europaea* (cultivated olive),
    82, 123
  O. *oleaster* (wild olive), 82
Olencira, 69
Orca gladiator (killer whale), 72
orchid hybrids, 89
Orchidaceae, 159
Orchis, 27
Origanum, 159

Plato, 14, 37, 146–8, 154, 232, 240
  and essentialism, 151, 200–1
Plato (spiders), 37
Plinia (myrtles), 165
Pliny the Elder, 18, 25, 162–5, 217
Plumier, Charles, 165
Poales (grasses), 159
Polemistus chewbacca (wasp), 30
Polemoniaceae (phloxes), 161
*Polichinellobizarrocomicburlescom-
  agicaraneus*, 75
Pollychisme, 62
Polygonaceae (rhubarb and
  sorrel), 162, 251
Polygonatum commutatum
  (solomon's seal), 179
Polygonia c-album (comma
  butterfly), 109
Polygonum raii (Ray's knotweed), 178
Polyporus squamosus (bracket
  fungus), xii
Pomacea canaliculata (channelled
  applesnail), 55
Pompholyxophrys punicea, 132
Pont, Adrian, 118
Porphyra umbilicalis, 115
Pragmatic Species Concept, 245
Pratchett, Terry, 31, 38
prefixes, 85–8
Priapulida, 28
Prideaux, Gavin, 136
Primula vulgaris (primrose), 21
priority, 130–4, 223
*Proceedings of the Zoological
  Society*, 46
Prokaryota, 86, 101
pronunciation, 62–7
protologues, 94
Prunella vulgaris (dunnock), 129
Prunella vulgaris (self-heal), 129

Psathyrella, 128
  P. ammophila, 44
Psephophorus terrypratchetti (fossil
  turtle), 38
Pseudoboletus parasiticus, 45
Pseudohydnum (jelly fungi), 86
Pterodactylus, 65
Ptinus (beetles), 50
Puccinia canaliculata, 55
Puffinus puffinus (Manx shear-
  water), 10

Qantassaurus, 21
Quercus robur (European oak), xiv, 11

Radde, Gustav, 80
Radicula, 109n
Rafinesque, Constantine Samuel,
  55, 134
Rana (frogs), 71
Ranunculus (buttercups), 159, 247
Raphanus (radishes), 159
  R. sativus (garden radish), 173–4
Ray, John, 18, 175–8, 215
Raymond-Hamet, 50
Rea, Carleton, 248–9
Renanthera (orchids), 89
Reptilia, 258–60
Rhea, 27
Rhinoceros unicornis, 113
Richardson, Mike, 92–6, 98
Richer de Belleval, Pierre, 76
Ridgeia piscesae (tubeworm), 240
ring species, 239
Rivinus, Augustus Quirinus, 80, 179
Rodentia, 17, 188n
Rolander, Daniel, 29, 32–3
Rosales, 174
Rotaovula hirohitoi, 37
Rothman, Johan, 182